U0190281

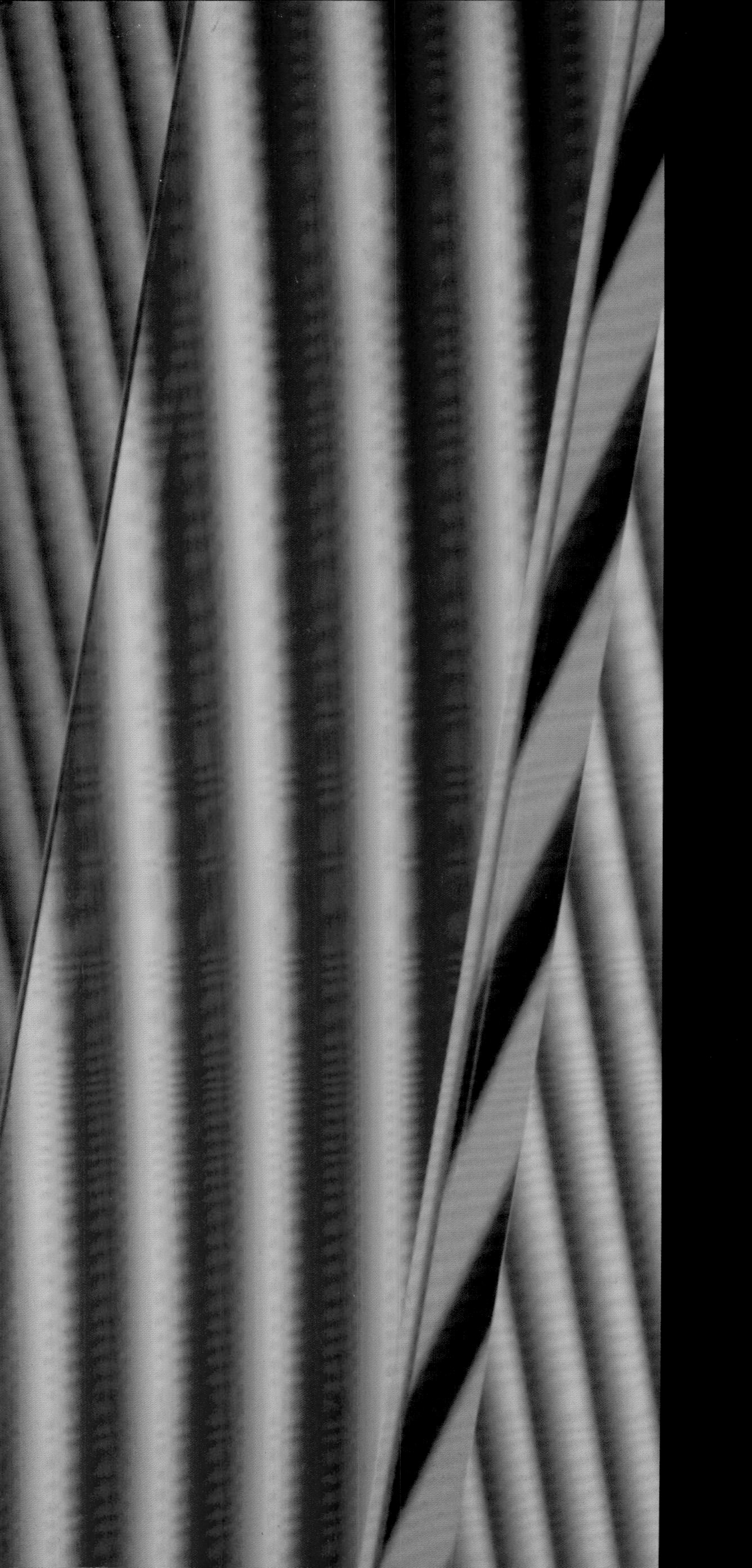

见微知著

纳米科学

〔美〕Felice C. Frankel
〔美〕George M. Whitesides 著
曾 杰|王 燚|秦 冬|夏幼南 译

No Small Matter

Science on the Nanoscale

中国科学技术大学出版社

内 容 简 介

人类所赖以生存的自然是极其广阔的。如果只凭借直觉,我们对这个世界的认识就只能局限于一个很窄的范围——介乎于"很大"和"很小"的"中间范围"。本书讲述的是微观世界的故事,它和经验有别,有时候甚至会完全超乎想象。事实上,"小"一直在吸引着我们:精致手表内的齿轮和弹簧、刺绣、蜘蛛网……每一样都很奇妙。我们既是制造精致手表的能工巧匠,又是懂得欣赏精美刺绣的鉴赏家。纳米技术已经成为信息社会不可或缺的一部分,并且这种重要性还在不断增加,可以说,纳米科技为我们的文明社会打开了通往未知世界的又一扇大门。本书所面向的对象是那些对科学和应用感兴趣的大众读者。从量子世界中有违直觉的规律,到生命和细胞进化的传奇,再到信息和能源技术以及医学……纳米科技所涉足的领域之广让人惊叹,这也给我们带来了极大的研究热情。同时,我们希望把这份热情分享给各行各业的读者,给热衷于科研事业的研究人员,给那些对新奇事物充满好奇心的高中生和老师,给任何一个有求知欲望的人。

安徽省版权局著作权合同登记号:第 12141380 号
NO SMALL MATTER: *SCIENCE ON THE NANOSCALE*, First Edition (ISBN 978-0-647-03566-9) by Felice C. Frankel & George M. Whitesides.

图书在版编目(CIP)数据

见微知著:纳米科学/(美)弗兰柯(Frankel, F.C.),(美)怀特塞兹(Whitesides, G.M.)著;曾杰等译. —合肥:中国科学技术大学出版社,2014.5 (2019.8 重印)

原书名:No Small Matter: Science on the Nanoscale

ISBN 978-7-312-03377-3

Ⅰ. 见… Ⅱ. ①弗…②怀…③曾… Ⅲ. 纳米技术 Ⅳ. TB383

中国版本图书馆 CIP 数据核字(2014)第 010959 号

出版 中国科学技术大学出版社
安徽省合肥市金寨路 96 号,230026
http://press.ustc.edu.cn
https://zgkxjsdxcbs.tmall.com
印刷 合肥宏基印刷有限公司
发行 中国科学技术大学出版社
经销 全国新华书店
开本 889 mm×1194 mm 1/12
印张 16
字数 498 千
版次 2014 年 5 月第 1 版
印次 2019 年 8 月第 2 次印刷
定价 99.00 元

序

蓝天白云、高峰深涧、鸟巢水立方……我们享受着宏观世界里这些触手可及的美好事物，却对存在于微观世界里同样的美景视而不见。你能想象出你的指纹有多么独特和神奇吗？你见过霸气的细胞，或者张牙舞爪的病毒吗？也许你从未意识到，这些和你朝夕相处的小东西也会成为一本写真集里面的主人公。本书给你提供一双见微知著的眼睛，让你能够从一个独特的角度去感受微观世界的奇妙，从而提供一次科学和艺术的完美邂逅。

原书作者 Felice C. Frankel 和 George M. Whitesides 为我们展现了一部精心编写的“微纳科技概述”，用美妙的图片引领我们走进未知的微观世界。你将在世界著名科学家和知名摄影师的带领下俯瞰微纳米科学的昨天、今天以及明天。Whitesides 教授是美国科学院、工程院、人文和科学学院三院院士，也是全球论文被引用次数最多的著名化学家，他在化学、材料科学以及近年来兴起的纳米科学等领域开创了众多研究方向。我本人于 1998 年在哈佛大学访问期间与他结识。当年，他送给我一块基于软刻技术设计的硅橡胶印章，可用来制作很多诸如这本书中所描绘的精美微小结构，至今我还收藏着这块印章。Frankel 是麻省理工学院的科研人员和科学摄影师，近二十年来她一直致力于用摄影等艺术手段来向人们展示高深莫测的科学发现。

作者用风趣幽默的语言深入浅出地向读者描绘出微纳米科学的神奇，并且巧妙地结合精美的图片来隐喻这些微小的事物之中所体现的科学之道。书中图片的获取使用了从拍摄单个细胞的光学显微镜到拍摄单个分子的扫描隧道显微镜等多种显微技术。这些精密的仪器或是借助绚烂多彩的可见光，或是依靠自由洒脱的电子，让我们可以从不同的角度去观察和体会微观世界的神奇。

Preface

In the macroscopic world, we can effortlessly enjoy many wonderful things around us, including beautiful clouds under a blue sky, breathtaking mountains and canyons, and landmarks in Beijing such as the Olympic Stadium and Aquatic Center. However, in the microscopic world, everything becomes invisible to our naked eyes. For example, can you imagine how unique and exquisite your fingerprints are? Have you ever caught invasive cells or unfriendly viruses in action? Perhaps you have never envisioned that these little things in and around us would one day become the role players in an album dedicated to the big science in a small world. With unrivaled clarity and stunning elegance, this book guides us to a narrative and visual voyage into the “micro-and nano-world.”

Felice C. Frankel and George M. Whitesides, the original authors of this book, meticulously crafted every piece of the book to provide portraits of the small world with astonishing pictures and to tell stories about big science in layman’s language. Under the guidance of a renowned photographer and a world-class scientist, you will quickly learn about the remarkable properties of “small stuff,” as well as their revolutionary applications and societal impacts. Whitesides is a member of the National Academy of Sciences and National Academy of Engineering, USA. As a pioneer in nanotechnology and many other areas, Whitesides is one of the most innovative and productive scientists in the world. I first met him in person during my visit to Harvard in 1998. He gave me an elastomeric stamp with patterned microstructures on the surface. Using a technique known as soft lithography, the stamp can be used to quickly generate small, complex structures, just like some of those in this book. After so many years, I still keep this stamp in my collection. Frankel is a research scientist and photographer at MIT. Over the past 20 years or so, she has devoted herself to the development of new visualization tools and methods for better understanding scientific concepts and discoveries.

The authors help us envision the invisible small things by thoughtfully selecting a large set of attractive and expressive photographs. Some of the pictures were taken using microscopy tools, including the optical micrographs of individual cells and scanning tunneling microscopy images of individual molecules. These tools are based on either the colorful visible light or the freely moving electrons, and the latter can also behave just like light waves. Only with the assistance of these microscopy techniques can we start to understand the essence of the micro-and nano-world. With the use of metaphors, the texts complement the pictures with lucid and narrative explanations about key concepts and fundamental principles in nanoscience that are otherwise very difficult to appreciate.

Imagine that some of the tiny structures described in this book will make your computer run faster while others will save your life when you are infected by bacteria or viruses. Can you wait to learn more about them? Let's start our journey to explore this wonderland!

Chunli Bai
President, Chinese Academy of Sciences
February 2014, in Beijing

试想一下,书中描绘的某些小东西将来可能会让你的电脑运行得更快,有些则会在你感染细菌或病毒时冲锋陷阵为你赢得生机。是不是很好奇它们的庐山真面目?让我们一起开始这次探索的旅程吧!

白春礼

中国科学院院长
2014年2月于北京

目 次

绪　论

人类所赖以生存的自然是极其广阔的。如果只凭借直觉，我们对这个世界的认识就只能局限于一个很窄的尺度范围——介于"很大"和"很小"之间。在这个范围内，我们可以感受到碎片扎入指甲时的刺痛，也能感受到一袋土豆的重量。从扔出一个垒球到抱起一匹小马驹，我们的身体都会自然而然地有所反应。但我们从没有扔出过像星系这么庞大的物体，也不曾坐在单个的原子上。可见，光靠直觉是无法理解世间万物的。我们可以利用生活赋予的经验来理解那些尺寸处于"中间范围"的物体，久而久之，经验就变成了"本能"。但是，当物体远远大于或者小于这一尺度范围时，我们就无法再用同样的方式去理解它们。

本书讲述的是微观世界的故事，它有别于经验，有时候甚至会完全超乎想象。事实上，"小"一直在吸引着我们：精致手表内的齿轮和弹簧、刺绣、蜘蛛……每一样都十分神奇。我们既是制造精致手表的能工巧匠，又是懂得欣赏精美刺绣的鉴赏家。

如今，我们对微小的事物特别感兴趣。为什么？为什么是现在？这是因为我们有了新的眼睛（即测量工具），可以观察那些以前无法观察的事物，同时又有了可以在原子尺度上雕琢材料的新工具。我们很好奇，那些确实存在却无论怎么努力也分辨不清的东西到底是什么。同时，我们也想知道，假如一个原子被挪了位置，结果又会如何。

我们生来就爱捣鼓。小的时候，我们总想把叉子戳进电插座来看看究竟会发生什么。如今，我们在纳米世界里同样也发现了许多富有神秘色彩的"插座"，好奇心驱使我们用各种"小叉子"去试探它们，这样做还会烧掉"保险丝"吗？在确保正常工作的前提下，插座可以小到什么程度？再小下去的话，又会如何？

小自有小的好处，有时候还表现得特别明显。我们总想开发出新的技术，用于制造微小的器件，有的放矢地为人类服务。譬如，制造计算机芯片、操纵分子、操控细胞等等。可以这么说，搭建小世界的能力代表着新的产品、就业机会，以及数不清的财富等。资本家们对其早已垂涎三尺了。

Introduction

WE UNDERSTAND THE WORLD INTUITIVELY over only a tiny range of sizes—sizes in the middle, between "very large" and "very small." We know how a splinter stings when it slips under our thumbnail, how much a sack of potatoes weighs, how it feels to throw a softball, how difficult it is to pick up a pony. We have never personally dropped a galaxy, or knowingly sat on an individual atom. When things are approximately our size, we understand them through repeated experience. Over time, our accumulated experience settles into intuition. When things are much bigger than we are, or much smaller, we don't *understand* them in the same sense.

This book is about small things. They're *different*—sometimes *really*, and enthrallingly, different. We humans have always been fascinated by "small": the gears and springs of a fine watch, embroidery, a jumping spider—each is a distinct kind of marvel. We think of ourselves as master artisans, and we have a connoisseur's appreciation of delicate work.

But small things are *now* particularly interesting to us. Why? Why *now*? To start, we have new "eyes"—new measuring devices—to see what has always before been invisible; also new tools to sculpt matter at the scale of atoms. We are curious to know: what *are* those things that we cannot see, no matter how hard we squint? We know they are there, but not what they look like. And by the way, what happens if we push *this* atom over *there*?

We humans are incorrigible tinkerers. From childhood on, we have stuck forks into light sockets to see what would happen. Now we have an entirely new set of tiny sockets into which we can stick tiny new forks. Can minuscule sockets still blow big fuses? How small can something *be*, we wonder, and still work? And if we make it even smaller ...?

Small things are, of course, sometimes also very valuable. We would like to *exploit* technologies that make the smallest things for our various ends: to fabricate the components of computer chips, to control the behavior of molecules, to manipulate cells. The ability to build very small things has the potential to generate new jobs, goods, and wealth. Capitalism licks its chops in contented, avaricious anticipation.

There is also the matter of "life." "Small" is a door to the microscopic palace we call the cell. We have finally realized that a master craftsman—evolution—has been at work long before us (in fact, long before we existed as a species), trying out the smallest designs. Among the most sophisticated nanomachines and nanodevices we know are the components of living cells.

And the cell itself—a micron-scale object—is complex in ways we do not begin to understand, and will not be able to mimic for many decades. Nature is an almost inexhaustible store of marvelous demonstrations of designs—the system for efficient packing of DNA, the sensors to communicate from outside to inside cells—that we might never have imagined. Nature also bristles with helpful hints about unfamiliar strategies for fabrication.

And, finally, our own evolution has blessed and burdened us with explorer's genes. We simply *must* know what is there, at the smallest scale. We can't help ourselves. We can no more ignore the nanoworld than Hillary could ignore Everest, Neil Armstrong the moon, or He Zheng the Indian Ocean.

The last few decades have seen the germination of new methods for seeing and manipulating minute things, and the flowering of science based on these methods. We happily anticipate continuing harvests of technologies and their products. But our skills in investigation and fabrication go only so far. What happens when things shrink beyond the range of our intuition and experience? Are small things just like big ones, only smaller? Or are they really *strange*? Despite a difference in size, an elephant shrew is, after all, the physiological first cousin to an elephant, and a seed pearl is not so different from a wrecking ball. But a nanotube and a soda straw are *fundamentally* different, as are the rotary motors that power bacteria and locomotives. One of the fascinations of small things is that they turn out to be so alien, in spite of superficial similarities in shape or function to larger, more familiar relatives. Discovering these differences at the smallest scale is wonderfully engrossing, and using them can change (and has changed) the world.

Two words are often associated with small things: *nano* and *micro*. Each is a unit of size; both are invisibly small. A nanometer is a billionth of a meter, or the size of a small molecule, which is ten times the size of the smallest atom, hydrogen. A micron, or micrometer, is much larger—a millionth of a meter, or a little less than the diameter of a red blood cell. Still, it is invisible to the unaided eye. The smallest objects we can see without a microscope—a hair, for example—are about a hundred times larger, or about 100 microns in diameter.

So, "micro" is already invisible: why focus on even smaller things? Why go beyond invisible, to nano? We offer eight reasons.

Information Technology. In the last fifty years, information technology (IT) has completely reshaped our global and local societies. Consumers see, play with, and talk to only a few of the most obvious devices—laptops, cell phones, electronic games, smart phones—that are based on nanoscale electronics and

细小的事物同样与生命息息相关。只有走进微小世界,才能打开细胞这座宫殿的大门。经过漫长的探索,我们终于认识到,有一个大师级的工匠——进化,早在我们存在之前就已经开始工作了。循序渐进,一点一点地尝试。我们现在知晓的最为复杂的纳米机器和纳米器件就是活细胞中的部件。其实,就连细胞本身这种微小尺度的物体,也复杂到让人难以理解,甚至在若干年内也没有办法去模拟它。大自然有用不完的奇妙构思,实现了诸如DNA的压缩包装、细胞内外之间的信息传输等这些我们难以想象的设计;同时它也留下了许多有用的线索,启发我们去探索新的加工手段。

人类的进化过程自然而然地赋予了我们探索的精神,这是祝福,也是负担。我们总想知道在最小的尺度上世界究竟是个什么样子。情不自禁的冲动,让我们再也无法忽视纳米世界的精彩,激励我们要像埃德蒙·希拉里登上珠穆朗玛峰、尼尔·阿姆斯特朗征服月球、郑和下西洋那样去探索与了解这个神秘的微小世界。

在过去的几十年里,我们见证了纳米科学与技术的蓬勃发展,包括用于观察和操控微小物体的各种新技术的诞生。我们可以乐观地预测,在这些技术进一步发展的同时,新的产品也会不断涌现。但这些探索的手段和工具最终会受到极限的挑战。人们不禁会问,如果将物体缩小到我们无法理解的尺度,会怎么样?小物体和大物体相比仅仅是尺寸更小,还是截然不同的"另一个"?如果不考虑"体型"上的差异,象鼩终究是大象的堂兄弟,生长珍珠的晶种与碎玻璃碴也没有本质区别。然而,碳纳米管和饮料吸管则截然不同,细菌的动力马达与火车头更是风马牛不相及的东西。原本我们所熟知的物体在变小之后,尽管从表面上看其与大物体在形状或功能上有相似之处,但本质截然不同。正因如此,探索微观尺度下的物理现象才这么令人着迷,并在潜移默化中改变着我们的世界。

提到微小物体时,我们通常会联想到两个词:"纳"和"微"。它们都是小到肉眼无法分辨的尺寸单位。1纳米是十亿分之一米,相当于一个分子的大小,是最小的原子——氢原子的十倍。1微米是百万分之一米,相对要大得多,略小于一个红细胞的直径,不过用肉眼依然没法分辨。在没有显微镜的情况下,肉眼最多只能分辨直径在100微米左右,如发丝般粗细的东西。

既然微米级的东西肉眼都难以分辨,干吗还要去关注更小的物体呢?为什么还要试图超越已经看不到的微米尺度,去研究纳米尺度的物质呢?我们有八大理由。

信息技术。在过去的五十年中,信息技术(IT)已经完全改变了我们的世界。我们完全依赖于笔记本电脑、手机、电子游戏机,以及智能手机来看、玩和交谈……这些常见的装置都基于纳米电子学和光学。事实上,还有更多不醒目的纳米尺度的装置,它们无处不在,就连它们的大部分用户和开发者也不清楚它们在何处。其中最重要的要数计算机(微处理器),现代社会几乎所有的工作都需要借助它来完成,如电话和空中交通管制网络的交换器,作战管理系统,全球大气监测传感器,电网、微波炉和钢铁厂的控制系统,医院里的病人监控设备,财务、人力资源业务处理设备和银行系统等等。几乎所有一切,从用手机致电尼日利亚商洽一筐香蕉的价钱到在社交网络中与人胡吹乱侃都依赖于电脑。

作为信息技术的核心,晶体管(微处理器的开关)中最小部分的尺寸以及晶体管间连接线的宽度,都已在纳米量级,远低于肉眼可分辨的尺度。晶体管的某些部分只有几纳米大小,其中的连接线仅有20纳米宽(比50个金原子首尾相接的长度还要短,比孩童睫毛的千分之一还要细)。令人难以置信的是,我们不仅学会了怎样去加工如此精细的结构,而且还知道如何大批量地生产。我们把计算机终端、计算机网络和连接不同计算机间的通信系统合称为因特网。这是人类目前最庞大、最野心勃勃也是最为错综复杂的工程。如今,因特网已经被推广到了全世界,并仍在不断扩展和复杂化,它就如同我们千辛万苦养育的孩子一样,我们并不知道它长大成人后会变成什么样,更无法判断它在未来会引领我们走向何方。

生命以及细胞。我们对包括自己在内的所有生命体及生命本身都有着浓厚的兴趣。而细胞是生命体的最小单位,所有的高等生物都由它们组成。当然,地球上数量最多的生物要数尺寸在1~10微米之间的单细胞生命体。

optics. The innumerable, less evident, devices are everywhere, and largely hidden to most of their human users and creators. The most important of these are the computers (microprocessors) that run just about everything in our modern world: the switches for the telephone and air traffic control networks; battle management systems; global atmospheric surveillance sensors; controllers for power grids, microwave ovens, and steel mills; devices for monitoring patients in hospitals, and for managing payrolls and accounts receivables, financial transfers, and the banking system. Everything—from a cell phone call that gets a good price on a cartload of bananas in Nigeria to the interactive babble and preening of social networking sites—depends on computers.

The smallest parts of transistors (the switches in microprocessors), and the widths of the wires that connect transistors together, are nanometer in scale—sizes far smaller than "invisible." Some parts of transistors are only a few nanometers in size, and the connecting wires are now shrinking past 20 nanometers—less than the length of 50 gold atoms laid end to end, thinner than one thousandth the width of a baby's eyelash. It is almost unbelievable that we have learned not just how to fabricate structures of such delicacy, but how to make them in unimaginable numbers. The resulting computers, and networks of computers, and communications systems that connect computers and computer-containing devices, make up the collective construction we call the Internet. It is the largest, most ambitious, most intricate engineering project our species has ever imagined. We have launched it into the world and are working diligently to help it grow in scale, complexity, and sophistication. But as with our animate children, we have no real idea what it will turn out to be when fully grown, or where it will lead us.

Life, and the Cell. We humans are intensely interested in everything else in our world that is alive, and in life itself. The smallest unit of life is the cell. Higher organisms are assemblies of them, but by far the most numerous of life forms with which we share the earth are single-cell organisms, which range from about 1 to 10 microns in size.

Cells are packed with complex, functional structures that are the products of the astonishing processes described by Darwin—aggregates of molecules selected from countless random variations because of the advantages they brought to the cell that had them. These aggregates replicate and repair DNA, transcribe DNA into RNA, make proteins and other functional parts of the cell, separate the contents of the cell into two equal parts during division, recognize pathogens, generate energy, sense and react to the environment, construct all the molecules required by the cell, move the cell and its components from place to place, and on and on. To understand life, we must understand the essential parts of the cell, and biological nanostructures are high on the list of those parts we must (but are only beginning to) understand.

Physical Measurement at the Atomic and Molecular Scale. Nano has brought a new capability to basic science: a way to "see" atoms and molecules on surfaces. One of the founding inventions of nanoscience is the set of instruments that enable, for the first time, the visualization of surface structures at the scale of atoms and molecules. These marvelous devices—scanning probe microscopes—operate on principles entirely unlike those of familiar optical microscopes.

The key element in every SPM is its cantilever—a tiny, flexible finger, usually with a tip that has been sharpened to a point a few atoms in radius. The cantilever and tip are the sensing element of the device—the functional equivalent of the lenses that focus and collect light in a microscope. This tip, when placed at a distance from the surface that is about the diameter of an atom, "feels" the proximity of the atoms on that surface as an attraction or repulsion, and it bends in response. Tiny though this movement may be, it can be quantified and used to generate an atomic-scale image of the surface. Scientific revolutions often ride on the backs of new tools, and so it has been with nanotechnology.

The Unique Properties of Small Pieces of Matter. Particles of matter with nanometer-scale dimensions have obvious differences from larger things. For one, a particle that is very small includes only a small number of atoms or molecules. Most of science deals with the average behaviors of very, very large numbers of molecules. (In the single shot of your afternoon espresso there are a million million million million molecules of water—technically 10^{24}—a number so large that it is literally unimaginable.) In a droplet of water with a radius of 2 nanometers—a droplet about the size of a small protein molecule—there will be about a thousand molecules of water. The statistics on which we so comfortably rely for accurate information about the average behaviors of large numbers of molecules begins to break down with nanoscale samples, and more of the idiosyncrasies of the individual molecules start to peek through. We know that thinking about the behaviors of the populations of cities is one thing; thinking about small groups in those cities is another; thinking about the individuals who live in a city is yet a third. The average income of all the people in a city, and the average income of a smaller group picked randomly out of its telephone book, and the income of a single, random individual are all different. So it is with molecules. Nanoscale collections are particularly interesting: they are neither "very

细胞装载着众多分子聚集体，或称之为生物纳米结构，它们有着令人瞠目结舌的复杂性和功能性。在达尔文笔下，这些聚集体在无数次随机性变异中，因其有益于细胞的生存，所以被保留了下来。它们复制和修复DNA，将DNA转换成RNA，生产蛋白质及细胞中的其他功能结构，在细胞分裂时将其平分，识别病原体，产生能量，感受环境并作出反应，构筑细胞所需要的所有分子，辅助细胞移动……想要理解生命的真谛，就必须先了解细胞中的纳米结构，目前我们还处于很初级的阶段。

原子分子尺度下的物理测量。纳米科技为探索自然科学带来了"看到"原子和分子的能力。其中的一项重要发明是一套可用来直接观察表面原子及分子结构的仪器，即扫描探针显微镜(SPM)，它开创了一个在操作原理上和传统光学显微镜完全不同的新时代。

扫描探针显微镜的关键部件是它的悬臂，其如同一根富有弹性的"手指"，前端还连有一个极小的探针。探针被削得尖尖的，顶端通常只有几个原子那么宽。悬臂和探针是扫描探针显微镜的关键部件，相当于光学显微镜中聚焦和收集光的透镜。当探针尖端距离待测表面仅一个原子直径的距离时，受到表面原子的吸引或排斥，探针尖端受力使悬臂发生弯曲，这种弯曲被光电探测器"感受"到后发出反馈，从而构建出待测表面原子结构的图像。科技革命通常依赖于新工具的诞生，纳米科技的革命也不外乎如此。

物质在微小尺度上的特性。纳米尺度的颗粒仅由少数的原子或者分子组成，它们和大物体之间有着显著的区别。大部分的科学现象来自众多分子的平均统计行为。一杯午后的咖啡中含有10^{24}个(一个大得令人难以想象的数字)水分子。一个半径为2纳米的水滴(相当于一个小蛋白质分子的尺寸)大概包含1000个水分子。我们通常依赖大量分子的平均行为及统计数据来总结规律，而这些规律并不适应于纳米尺度的材料，同时，单分子的特性在这个小尺度下开始显现。我们知道，城市全体居民或一些社团的集体行为和居民的个人行为是截然不同的。居民的平均收入、从电话本里随机选取的一群人的平均收入和随便一个人的收入也截然不同。对于分子，亦是如此。

纳米尺度上的物质特别有趣：它们既不是大量分子的集合体也并非单个分子，而是介于两者之间的一个特例。

纳米颗粒是由表面分子主宰的。当颗粒的直径小到只有几个分子大小，所有的这些分子就都会分布在表面或者相当靠近表面的位置。颗粒表面的分子和内部的分子有着完全不同的性质，主要是因为它们处在两个完全不同世界的交界处即异相界面上。表面分子那些有别于内部分子的特性通常十分有用，例如，铂的纳米颗粒可用作催化剂将原油转化成汽油。

材料。钢铁、混凝土、聚乙烯、玻璃、半导体硅、液晶、花岗岩、木头、浓缩铀等是我们用来加工电脑芯片、机翼、隐形眼镜、刀具、穿甲弹、装汽水的塑料瓶、发动机以及更多其他物品的基本元素。我们知道这些材料的原子结构，并且了解它们的宏观组成、性质以及功能。

但是在原子和器件之间存在着一个未知的纳米世界，涉及各种各样的纳米结构，例如金属的晶界、裂纹的尖端、聚合物中的结晶区、空洞、界面等等。这些不起眼的结构有时反而会对产品的功能起到决定性的作用。过去我们并没有能力对这些结构进行检测，而现在我们起码有了一些测量途径。同时，我们如今还能制备出一些具有新奇特性的材料，例如量子点（纳米世界中闪亮多彩的指示灯）、自组装单分子膜（覆盖表面的单层分子膜），以及精致得难以想象的纳米棒和纳米盒等。

量子现象。我们生活在一个与人类尺寸相当的世界里，原以为已经理解了支配这个世界的物理规律，例如球会自然下落，两个闭合的圆环不可能在靠近的时候穿过对方嵌套在一起，一个人不可能在两个地方同时出现，咖啡和咖啡伴侣在杯中会混在一起……但是当物质结构小到纳米尺度时，量子现象便会显现。电子、原子和分子所遵循的规律实在让人费解。虽然我们可以用所谓的量子力学来精确地描述这些规律，但是在很多情况下，其结论总是有悖于我们从日常生活中学到的东西，哪怕这些东西源自于痛苦的经历。比如说，水泥块总会砸伤我们的脚，我们不可能穿过一扇没打开的玻璃门……

这真有点让人摸不着头脑：一个熟悉的世界，它遵循可以理解的规律，然而组成它的原子所含的电子

large" numbers of molecules nor "single" molecules, but are rather at the border between the two.

Nanoscale particles are also "all surface." If a particle is small enough—only a few molecules across—then all of those molecules are close to a surface. Molecules at a surface have different properties from those in an interior: they exist at the heterogeneous interface between two more homogeneous worlds. The oddities and unusual properties of matter at surfaces can often be very useful (for example, when nanoscale particles of platinum are used as catalysts to convert crude oil into gasoline).

Materials. Steel, concrete, polyethylene, glass, silicon, liquid crystals, granite, wood, isotopically enriched uranium, and innumerable other materials are the structural and functional elements we use to build things: the chips for computers, the wings of airplanes, contact lenses, knives, armor penetrators, plastic bottles for fizzy water, engine blocks, and countless others. We understand the atomic structure of these materials, and we also understand their macroscopic forms and properties and functions.

But in between—between atoms and machines—lies a world of nanoscale structures that we understand less well: grain boundaries in metals, the tips of cracks, crystalline regions in polymers, voids, interfaces. Many of these structural elements are critically important in determining function, but we have not been able to examine them in the past. Now we can, or at least some of them. We can also make *new* materials with *new* properties: quantum dots (brilliant, multicolor beacons for the nanoworld); self-assembled monolayers (skins one molecule thick to cover surfaces); nanosized rods and boxes of incredible delicacy.

Quantum Phenomena. We live in a human-scale world governed by physical laws that we believe we understand: balls fall down, objects cannot pass through one another, a person cannot be in two places at once, coffee and cream mix in the cup. But as structures become small enough to reach nanometer scale, quantum phenomena begin to emerge. For quantum objects (if, in fact, we can speak of an electron or even a molecule as an object), the rules are completely and disorientingly different. We have excellent mathematical descriptions of these rules—we call them *quantum mechanics*—but what they say disobeys most of the lessons we humans have so painfully learned by dropping many concrete blocks on many toes, and by trying to walk through many glass doors without opening them.

It's a disorienting situation: a familiar world, whose structures follow understandable rules, emerging from a quantum world of electrons and nuclei—of atoms, molecules, and materials—that follow completely different, often counterintuitive rules. Nano is the range of sizes where this transition—from Alice in Wonderland to concrete blocks and window glass—occurs. How does it happen? How can we understand our reality, when it rests on components that exist in a different reality? We know the answer, mathematically; but "knowing" is not "understanding," and "calculation" is not "intuition." Nanoscience may help us understand how "quantum" blurs into "ordinary." And the study of behavior at the quantum level may also offer opportunities for new technologies—technologies so radically outside our experience that we may not even recognize them when they first appear.

Medicine. Since each of us is a nation of cells, and our cells are cities of nanostructures, it seems plausible that nanoscience and nanostructures should be useful in medicine—in illuminating the operating instructions of the cell, in selecting specific cells for destruction or protection, in on-time, precise delivery of drugs to tissues. So far, nanoscience has not produced a bonanza of new medical capabilities, but there are promising beginnings: magnetic colloids that improve the ability of magnetic resonance imaging to detect tumors; drugs dispersed as nanoparticles to improve their availability in the body.

The match in size between the nanostructures in cells and the tools of nanoscience seems so opportune that enthusiasm for the potential of "nanomedicine" only grows with time. We believe, optimistically, that it's just a matter of "more": more time, more money, more luck, and the application of more intelligence. Of course, we believe, optimistically, that *all* new technologies have unlimited promise. We'll see.

Energy, Water, and the Environment. Nanostructures are everywhere in the technologies used to produce and conserve energy. All liquid fuels (gasoline, diesel, jet fuel) start with crude oil—often black semisolids that would be useless in a gas tank. We rely on reactive metal nanostructures (catalysts) to convert them into conveniently pumpable, easily burnable liquids. Fuel cells could not work without nanoscale catalyst particles and nanoporous membranes. Membranes are also essential to the production of potable water from sea water by reverse osmosis.

Nanostructures are a ubiquitous part of the earth's natural climate control. The spray of seawater at the beach breaks up into droplets; when the water evaporates, nanoscale particles of salt remain. These in turn are important in forming clouds, which influence the heat balance of the planet. Dust, of course, has many sources: wind blowing over dry soil; fires; volcanoes; and others. They all contribute. Nanoscience offers the possibility that understanding these sorts of natural processes might, at best, lead to new skills use-

和原子核却遵循着完全不同的规律。纳米则是介于两者之间的尺寸范围，一边是爱丽丝的奇境，另一边则是我们身处的现实世界。这是如何发生的呢？该如何理解我们所熟知的这个世界其实是由另一个完全不同的世界所构成的事实呢？从数学角度来看，我们或许知道答案，但知道并不等于理解，计算的结果并不等于我们的直觉。或许，纳米科学可以帮助我们理解"不寻常的量子物体"怎么变成了"寻常的宏观世界"。针对量子效应的研究为很多新技术的诞生打下了坚实的基础，即便如此，这些新技术由于超越我们的认知经验，在其刚刚诞生时也遭受了很多质疑。

医学。如果说人体是细胞构成的王国，那么细胞就是用纳米结构组建的城市。由此可断定，纳米科技的发展必定会有益于医学研究，尤其是在阐释细胞的操作指令、选择性地破坏或保护特定细胞、精确输送药物到达特定细胞组织等方面。目前，虽然纳米科技并没有催生出很多具备新性能的药物，但是其作用不可忽视，例如，磁性纳米材料可以提高用于检测肿瘤的磁共振成像的分辨率，服用纳米颗粒形式的药物有助于提高其被身体吸收的效率，等等。

纳米科技的研究手段与细胞内的纳米结构在尺度上的匹配性为我们提供了一个很好的研究平台，使得我们对"纳米医学"的研究热情与日俱增。我们乐观地相信，纳米科技与现代医学技术的结合必定会惠及天下，我们所需要的只是投入更多的时间、经费、人力和物力。我们也坚信，技术无尽头，探索无止境。

能源、水和环境。纳米结构在能源工业中的应用随处可见。所有的液体燃料（包括汽油、柴油和航空煤油）都是从原油中提取的，原油通常是黑色的黏稠液体，不能直接用作交通工具的燃料。我们必须利用金属纳米颗粒催化反应将其转变为更容易泵取与燃烧的液体。此外，燃料电池也需要纳米颗粒催化剂和纳米孔径的薄膜才能工作。多孔膜通过反渗透法也可用来净化海水。

纳米结构也参与了全球气候的自然控制。海水拍打堤岸溅射出一粒粒小液滴，随着水分的逐渐蒸发，便在空气中形成盐的纳米颗粒；风化的岩石、燃烧的山林、喷发的火山都会产生微小的尘埃，飘向空中。这些颗粒和尘埃都对云的形成起着重要作用，继而影响着地球的热量平衡。在纳米尺度下解析这些自然现象至少有助于我们理解恶劣环境产生的根源，甚至发展出控制和改善环境的新技术。

如果说对“微小”世界的探索也有新旧之分，那这条界线就在纳米和微米之间。由少量原子和分子构成的纳米世界里充满新奇与诱惑，而我们对于新鲜事物又总是充满热情，因此，纳米尺度的物体就自然而然地成为了现代自然科学研究的宠儿。然而，研究微米这个“旧”世界仍然具有很大的价值，我们今天在纳米世界中所取得的硕果也恰恰是建立在对微米体系的研究基础之上。在过去的几十年中，包括信息领域的电子和光学技术（操控电子和光子运动的技术）在内的那些给世界带来巨大变化的高科技，都是基于在微米尺度下调控物质结构而发展起来的。在未来的几十年中，微米尺寸的物质和结构仍然会是各种装置与器件中最核心的组成部分。

从宏观到微米再到纳米世界的无畏探索，必定会给我们带来意想不到的收获。与此同时，我们也可能会得到不尽如人意的结果，甚至会遭遇难以预测的危险。我们知道纳米结构与宏观结构有着不同的性质，我们也知道在认识纳米世界的独特性时，直觉感受并不完全可靠。纳米世界的某些特性有时会给那些勇敢的探索者带来危险，甚至会给无辜的旁观者带来灾祸。纳米技术的飞速发展，就如同在科学进程的大汤池中加几勺调味料一样，或许不会改变这锅汤原有的味道，却也无法确定它的安全性。任何新的事物都有无法预见的危险，同样也伴随着无法衡量的利益，最明智的选择就是要对两者都做好心理准备。到目前为止，虽然作为新兴技术的纳米科技还没有出现特别令人惴惴不安的惨象，但我们也必须小心谨慎、步步为营。

纳米技术已经成为信息社会不可缺少的一部分，并且这种重要性还在不断增加，可以说，纳米科技为我们的文明社会打开了又一扇通往未知世界的大门。这本书所面向的对象是那些对科学和应用感兴趣的大众读者。从量子世界中有违直觉的规律，到生命和细胞进化的传奇，再到信息和能源技术以及医学……纳米科技所涉足的领域之广让人惊叹，这也极大地激发了我们的研究热情。同时我们希望把这份热情分享

ful in controlling our environment (we hope by design, and to our benefit) and, at worst, help us comprehend what went wrong.

The border between *nano* and *micro* is the border between the new and the old in the science of small things. Nanoscale structures—objects made of small numbers of atoms and molecules—are one of the greenest pastures in modern science. In our enthusiasm for the new, however, we should not forget that we have grown fat on the old. Nano is sexy, but micro still pays many of the bills. Most of the technologies that have changed the world in the last decades—especially in electronics and photonics (technologies that manipulate information as puffs of electrons or blinks of light)—were developed using micron-scale structures; and for decades to come, micro will be an essential part of the swarms of devices that populate our world.

In venturing from macro to micro to nano, we must watch for unintended consequences. As with all exploration, and all new technologies, there is the possibility of nasty mistakes and unanticipated dangers. We realize that nanostructures have different properties from macrostructures; we know that our intuition is not a perfect guide to their peculiarities. Perhaps some of these properties pose a hazard to those who explore the nanoworld, and even to innocent bystanders. In the cottage industry that has grown up to divine the perils in nanotechnology, a trace of apocalypse adds spice to what may be a fairly ordinary scientific soup. But there *are* always unforeseen risks, as well as unforeseen benefits, with any new activity. It is prudent to try to anticipate both. Nothing particularly worrisome has emerged from nanoscience so far, but as more new technologies emerge, we should examine them carefully, with caution in mind.

So: nanoscience is an alluring new starting point for our civilization as it explores the world in which we live; nanotechnology has already become an essential part of the nervous system of a global society that depends on information to keep its parts working cooperatively (and will depend on it even more in the future). For whom, then, have we—the authors—constructed this account of the nanoworld? This book is intended for general readers who are interested in the implications and applications of science. We are ourselves in love with the subject of nanoscience, in part because of its range—from the intellectual disequilibria of the quantum world, through the evolving saga of life and the cell, to the technologies of information and energy and medicine. We would be delighted to share our enthusiasm with those who read the science pages of a newspaper; with investigators who enjoy science outside their specialty; with high school students and teachers curious about cool stuff; and with just about anyone fascinated with how the world works.

We have used both images and prose to give a sense of the many things that are too small for the eye to see. Images of the invisible world pose a particular problem because we usually cannot generate them with photography. Some of the images we show may *look* like photographs, but they were not produced using light for illumination. They represent the mating of new scientific instruments that can examine small things—scanning probe microscopes, electron microscopes—with computers and computer graphics. A light microscope is simply a magnifying lens that shows what our eye could see if it had the resolution. An electron microscope does not see light at all, but rather electrons (our own eyes are entirely blind to electrons). And an atomic force microscope (a particular kind of scanning probe microscope) produces images that resemble photographs but does not "see" in any sense. Rather, it "feels" the surface and feeds data back to a computer, which cleverly creates maps and converts them into visible images that *look* like photographs.

Images of the very unfamiliar phenomena common in nanoscience—especially quantum behavior—are even more problematic to produce, since these phenomena often cannot be understood fully except through mathematics. Fortunately, an image need not be a photograph of the thing itself to be useful. Just as a hand, a candle, and a wall can create a shadow that gives some sense of a rabbit's shape, the images we show give some idea of the shape of molecules and their arrangement into materials and cells. We also freely use analogies and metaphors—both in images and in words—based on familiar objects; they help almost everyone understand new ideas. We rely on them shamelessly—although we know they are not exact—to try to illuminate unexpected properties of nanosystems. When they are simply incapable of explaining precisely what is going on—as is often the problem in discussing quantum behaviors—analogies and metaphors are still much better than nothing. They give *some* sense of the origins of the unexpected properties of small things, and, failing even in that, they let us express our amazement (and sometimes incomprehension) more clearly than just blurting out "Sorry, we can't explain it, and actually we don't entirely get it ourselves!"

The difference between analogy and reality is usually obvious in this book. We also make an effort to distinguish between various types of images—which "photographs" involve light and "seeing" in the conventional sense, and which images represent the ruminations of a computer trying to summarize science for a species equipped with a specialized set of sensory organs called "eyes." Still, the distinctions may not, after all, be so important. The point is to get a sense for the shape of the thing, not to fret about the details of

给各行各业的读者,比如说那些经常翻阅报纸科技版的读者,热衷于科研并对其他领域感兴趣的研究人员,那些对新奇事物充满好奇心的高中生和老师,或任何一个有求知欲望的人。

在本书中,我们用图像和文字共同去描述那些小到肉眼分辨不清的东西。由于不能直接用普通相机拍摄到它们的照片,这对用图像来描绘它们就有了一定的困难。有些图像虽然看上去很像真实的照片,但实际上它们并不是通过光学成像得到的,而是借助扫描探针显微镜、电子显微镜等可以"看到"小物体的新仪器、电脑以及绘图软件配合完成的。光学显微镜的工作原理是在有限的分辨率范围内简单地用透镜把东西放大到我们肉眼可以看到的程度。电子显微镜不是用光而是用我们的眼睛完全看不到的电子来"看"。原子力显微镜采集的图像和普通照片很像,但是跟"拍照"毫无关联。事实上,它是通过探针"感触"物质表面,将数据传回给电脑,智能地模拟出表面起伏程度并将其转换成看起来很像照片的图像。

纳米科学中有很多我们不熟悉的现象,特别是量子行为。这些行为除了通过数学手段很难被完全理解,因此,它们的直观图像就更难得到了。所幸,我们在描绘事物的时候并不一定非要使用真实的照片,就像在描绘兔子形状的时候,我们也完全可以用手、蜡烛和墙创造一个逼真的影子。与此类似,我们在书中也使用了一些可以演示分子形状以及它们在材料和细胞内部排列的图像。我们还尝试了在图像和文字中灵活运用类比和比喻,用它们来共同阐释纳米系统的特殊性质。尽管知道这样的做法可能欠推敲,但这不失为一种可以帮助我们理解新事物或新构想的有效方法。当我们没办法通过简洁的语言来解释清楚眼前发生的事情时,类比和比喻就显得格外重要了,它们让我们对那些小事物的特殊性能有初步的印象,也让我们能清晰地表达出自己的观点和心情,而不是无奈地说:"对不起,我们解释不清楚,准确地说连我们自己也没有完全弄明白。"

本书中运用的类比在很多情况下和现实还是有差异的。我们努力将两类图像分开,一类是借助光拍摄的普通照片,一类是通过仪器和计算机对自然现象模拟出来的图像。也许它们之间的区别并没有那么重要，重要的是我们需要对事物的本质有比较清楚的认识。再者，即使是我们的肉眼直接看到的世界，也是进化过程中产生的“艺术品”。举个例子，小鸟和蜜蜂对色彩的解读就是人类永远无法企及的。

最后，我们如何评价“纳米”呢？首先，纳米科技才刚刚开始。其次，就像科学的每一次飞跃一样，纳米也正以一种独特的方式展示它的魅力。量子力学是一个独立的、全新的概念，虽然它在物理学界掀起的轩然大波非常短暂，但是对这个概念的进一步研究已经持续了一个多世纪。基因组学是个百宝箱，它可以用来读取基因组的信息、按特定顺序将基因排列等，这些新技术已经改写了细胞生物学，也让我们对生命有了全新的认识。化学的发展是一小步一小步走过来的，用催化反应来生产燃料、材料以及药物，发明刻画复杂分子结构的仪器，发展复杂的合成方法，以及把理论用于理解大量分子平均行为，这些都推动着我们的社会朝着更为舒适和健康的方向发展。我们知道，天文学依赖新的望远镜去探究时间的起源；而对于材料科学，这个学科之所以存在是因为政府需要新材料来发展像飞机那样复杂的机器。

我们还无法预见纳米科技最终会对未来产生怎样的影响，但是我们知道它正沿着自己的轨迹前进。它的缔造者有发明计算机和网络的日本电气、英特尔、谷歌以及华为的工程师，有发明测量和解释量子行为的物理学家，有开展合成、催化、加工的化学和材料学家，有探索生命之光的生物学家……纳米的世界已经像集市一样热闹，它是许多领域的集合，而不是一个全新的个体。无论如何，它正在转换我们的思维方式，改变科学与工程学之间的合作关系，重塑技术生态，同时它又促进不同领域间的相互了解与沟通。

就目前已产生的影响来看，它已经非常之酷，也非常有用。

how we perceive it. In any event, what we see, even with light, is an artifact of our evolution: birds and bees perceive colors that human eyes never will.

And at the end, what can we say about nano? First, that it is *not* at the end, it is at the beginning. Second, it is—as is every large-scale development in science—unique in the way it is unfolding. Quantum mechanics was a single, radically new idea, which exploded over a very short time in the world of physics; the reverberation of this explosion has kept much of science going for a century. Genomics was a toolbox—containing the implements needed to read the information in the genome and rebuild it to order—whose applications have transformed cell biology and our understanding of life. Chemistry developed by taking many smaller steps—catalytic production of fuels and materials and drugs, development of instruments to characterize the structures of the most complex molecules, invention of sophisticated methods of synthesis, construction of a theory to understand the inevitabilities of large numbers of molecules—that have moved society in the directions of comfort and health. Astronomy has relied on new telescopes to see to the beginning of time as we know it. Materials science was created by the government to solve problems in complex machines such as airplanes.

We will not know the ultimate impact of nanoscience and nanotechnology for many years. We *do* know that it is following its own path. Its creators have been the most variegated that a discipline (if it can be called a discipline) has seen: engineers at NEC and Intel, Google and Huawei, building computers and the Internet; physicists providing tools for measurement and interpreting quantum behaviors; chemists and materials scientists synthesizing and catalyzing and fabricating; biologists phrasing the problems posed by "life." Nano has started as a village and a *suq*, with much buying and selling. It may turn out to be the integration of parts of many fields, rather than an entirely new concept. No matter. It is transforming the way we think of our world; it is changing the way science and engineering work together, and rewiring the sociology of technology; it is brokering the movement of knowledge among communities that barely knew of one another's existence.

Also, it's already very, very cool, and very useful.

SMALL

Small is different. Very small is *very* different.

Intuition and experience work with billiard balls and babies. We understand them, we think. Only mathematics works with electrons and atoms; they defy everyday intuition. And yet ... balls and babies are made of electrons and atoms. How does the commonplace emerge from the unimaginable?

When things are large, they are what they are. When they are small, it's a different game: they are what our measurements make them. At the marches—the regions too complicated for mathematics and too unfamiliar for intuition—strange creatures (with stranger behaviors) emerge ... and our "experience" turns out to be a Classic Comic version of reality (whatever reality might be).

小之道

微观尺度的事物很特别，越微小，越与众不同。

通过练习，我们可以很好地控制台球的运行轨迹，以及掌握婴儿的作息规律，这让我们对自己的直觉和经验充满信心。然而，电子和原子却颠覆了一切日常的经验和规律，唯有数学推演才能解释它们的行为。人们不禁会想，台球与婴儿也是由电子和原子构成的，那为什么这些平日里无法预知的东西能构成我们日常熟知的事物呢？

当事物足够大时，你看到的可能还是真实的它。然而当它变小时，一切都变了，这家伙的形态和位置都将随着我们的测量而千变万化。在这样一个被颠覆的世界里，一切都是那么的复杂和陌生，一些行为怪异的小家伙在其中横行无忌，数学在遇到它们之后就变得异常复杂，直觉又是那么的荒谬可笑。好吧，不管现实到底是什么样的，至少我们曾经的“金科玉律”已经被证明只不过是“一派胡言”。

Santa Maria

探索新大陆

1

Why sail off the edge of the earth? Why risk the terrible storms, and the more terrible sea monsters? Why did he choose to go?

For the usual reasons: the enticing prospect of wealth, the even more enticing prospect of fame; for curiosity and the thrill of gambling and one-upmanship. From restlessness and ruthlessness and boredom and a craving for admiration. Because he thought that he might, possibly, get away with it one more time. Columbus set off westward from Spain, aiming for the Indies. It was the wrong direction, but exploration was the right idea, and he found something more valuable than silk: an unclaimed continent!

Nanoscience may seem tame by comparison with exploration of uncharted oceans. It is not. True, an error does not make you snack-food for a sea monster. But what you learn! That the comfortable reality we know—the world where a book on the table is solidly on the table, and doors keep children safe inside—emerges from a *most* uncomfortable, deeper reality where things can be many places at once, and doors provide almost no barrier. We can probably never understand this shadowy, disorienting landscape intuitively, but mathematics guides us through it well.

And what do we discover? That electrons skittering through wires, and collecting in pools, change our society more than spices or silk. That life—the rose in the garden, the aphid on the rose, the microbes in the aphid's gut—is so astonishing that one forgets to breathe. That we just skim the surface of reality; we don't live there.

Columbus's first voyage found what is now called the Bahamas; only on the fourth, after many ricochets, did he accidentally hit Central America. He had no real idea of what he had found, or that others had been there before him. He could not foresee that the colonists he left behind would die or mutiny. But no matter. Columbus was in roughly the right place at roughly the right time. With nano, it is still unclear what is archipelago and what is continent, if, indeed, there is one. No matter: *small* is the right idea for our world and time.

The science of small things is spinning the web—not of silk, but of information—that has made silk roads obsolete. It helps us understand what it means to say that a silkworm is *alive*. It will help to generate the energy and water (and who knows what else?) whose value so far exceeds that of silk and diamonds.

以可怕的风暴为帆、海怪的鲜血为道、坚强的意志为舵，他出发了，去寻找世界的尽头。

为何要扬帆远航？对于哥伦布来说，这也许只是为了满足自己的好奇心，抑或是为了赢得一场豪赌，当然也可能是为了追逐名利和财富。刚开始，人们对哥伦布的探险之旅感到惶恐不安，甚至横眉冷对、无情抵制，但最终还是钦佩不已。曾经也许困难重重，也许命悬一线，但好在哥伦布始终坚信自己能够成功。他从西班牙的西海岸出发前往东印度，虽然选择了错误的航向，但仍然证明了探索未知的正确性。他发现了比丝绸更有价值的东西：一个全新的大陆！

相比未知的海域，探索纳米科学的旅程似乎安全多了，最起码你不会因为一个错误而成为海怪的美餐。但是你的发现会让你目瞪口呆、难以置信：在我们已经生活并适应了很久的世界里，放在桌上的书不会不翼而飞，大门也总能把孩子们安全地保护起来……然而，纳米世界完全是另外一番情景。在那里，桌上的书可以同时在几个地方出现，而所谓的大门也形同虚设。也许我们还不能一眼看透纳米这个混沌的世界，但是，感谢上天赐予的数学利剑，助我们披荆斩棘、开拓出一条光明之路。

那么，我们的探索之旅到底会发现什么呢？当电子流过电线再汇聚时，它们给社会带来的进步，远远超越东方的香料和丝绸。又比如说，生命的绚丽多彩，像玫瑰在花园里绽放，蚜虫在玫瑰上小憩，以及微生物在蚜虫体内喧闹，所有这些都使我们忘掉呼吸。当然，我们的探索往往是浅尝辄止、匆匆掠过，离真理还很远。

哥伦布第一次远航发现了现在被称为巴哈马群岛的地方。功夫不负有心人，经过多年的探索，他终于在第四次的航行中偶然发现了中美洲。当时，他并不知道他发现的是什么，也不知道别人在他之前已来过这里。他也没能预见自己所发现的这片土地日后会变成殖民地，并会经历暴乱和死亡。但这些都不重要，重要的是哥伦布在正确的时间来到了正确的地方。对纳米世界来说，它的"群岛"在何处，它的"大陆"又在何方，我们至今仍不得而知。但是只要存在，我们就总有一天能找到它们。不管怎么说，向更小的世界进发肯定是正确的决定。

微小物体的科学孕育了信息网络的诞生。它用信息代替蚕丝，编织了一个令丝绸之路黯然失色的网络。这个科学还帮助我们理解"蚕蛹"还活着的真正含义。此外，它还可以产生能量和水，说不定还会有其他更好的东西，这些东西的价值将远远超过丝绸和钻石。

Die gantze Welt in einem Kleberblat/Welches ist der Stadt Hannouer meines lieben Vaterlandes Wapen.

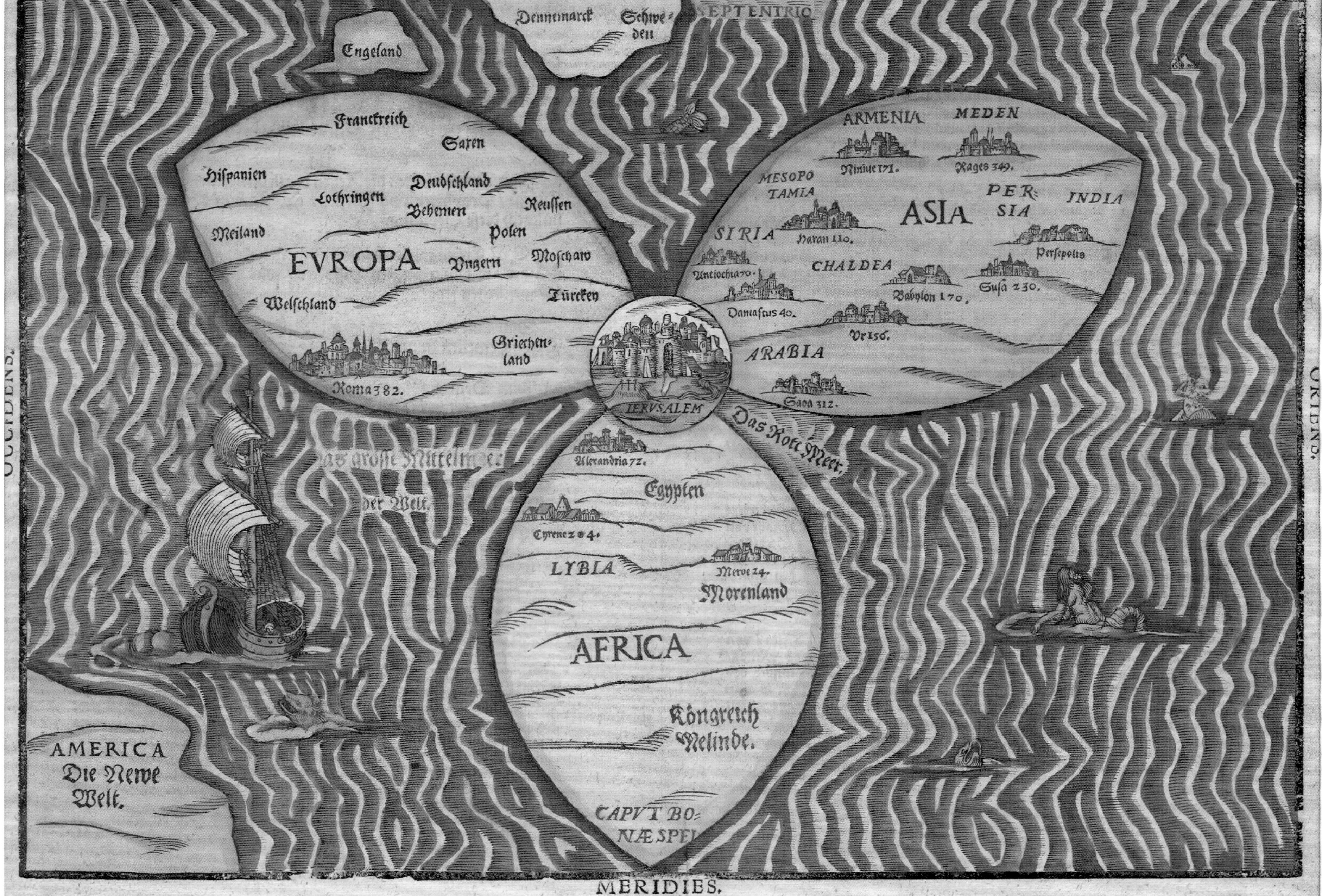

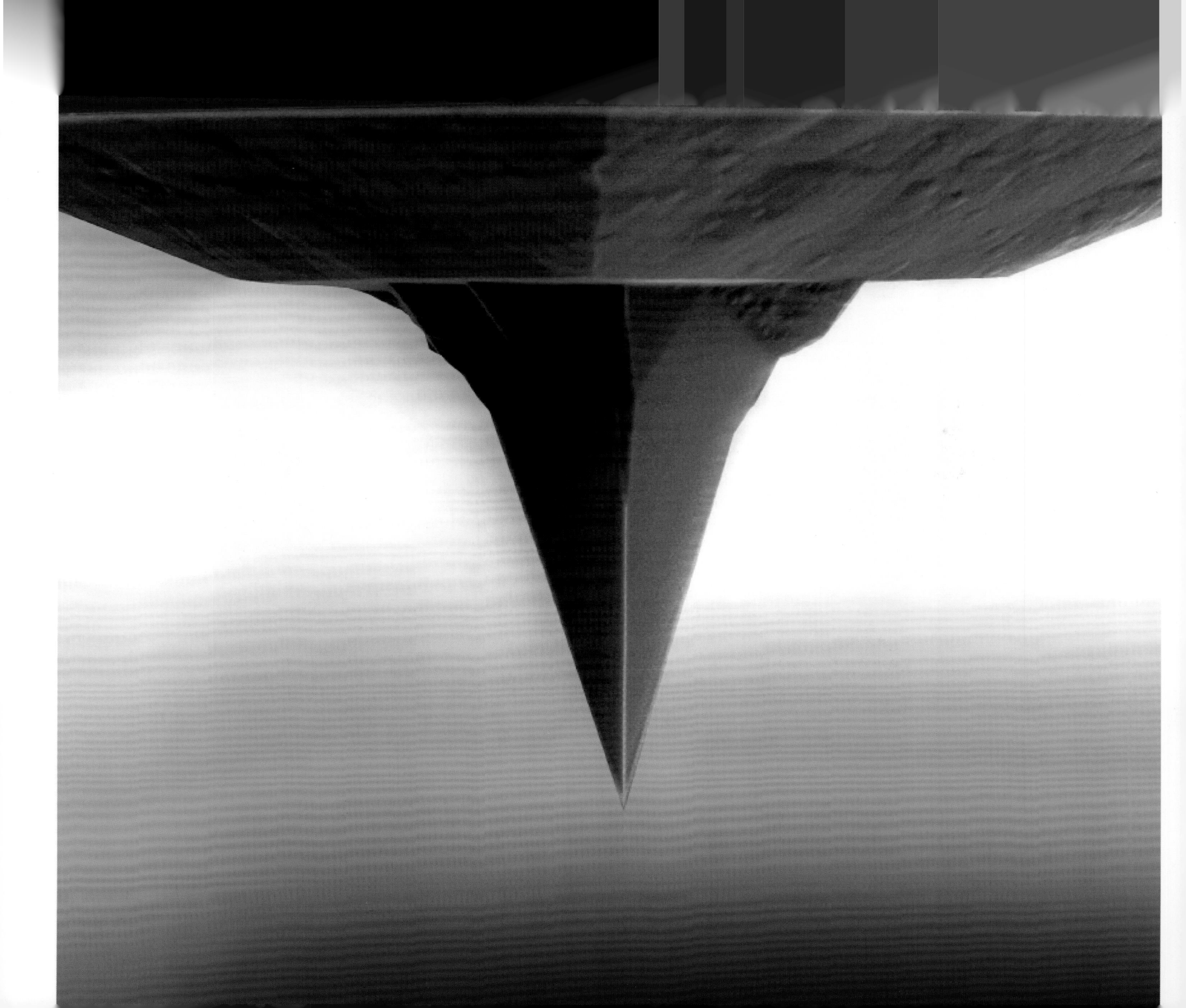

盲人摸象?

若没有放大镜,我们便很难明察秋毫,而光学显微镜则可以帮助我们看到更小的物体。可是,当物体的尺寸只有几百纳米时,由于光衍射,物体尖锐的边缘会产生阴影,从而使图像变得模糊不清。当然,我们可以使用X射线以及电子束来拍摄这些细微东西,哪怕它们小到只有几纳米的尺寸,但是我们不能再说"看"到了它们。对于真正的微观世界,我们就像得了眼翳。在这种情况下,我们如何探索纳米世界呢?

我们不妨向蠕虫祖先学习,用触觉代替视觉。蠕虫生活在黑暗的泥土中,眼睛就如同聋子的耳朵一样,只是个摆设而已,但它们可以用皮肤来感知周围的世界。在物竞天择中,用触觉探知没有光亮的世界是一个行之有效的方法。就好像猫和鲶鱼用它们的胡须来感知,而蟑螂则通过触须。

原子力显微镜(AFM)可以说是纳米科学的奠基石,它通过一根探针在物体的表面上扫描,精确地触摸到每一个原子。借助它,我们能够描绘出分子的形状,这足以让那些致力于分子研究却从没真正看到过分子的化学家们为之痴狂。

当然,原子力显微镜的分辨能力与探针的大小密切相关。这并不难理解,想象一下带着滑雪手套穿针引线时力不从心的情景吧。为了分辨到原子,探针的针尖与原子的尺寸必须旗鼓相当。图中的针尖看起来像是一座倒放的山峰,但实际上它是尖端被雕琢成只有几个原子宽的硅碎片。为了拍摄这幅图,我们用电子束来照亮这个针尖。

为了用触觉描绘出更复杂的图案,其中一个有效的方法就是采用多根针尖。其工作原理类似下页所示的儿童玩具:一个布满了可上下滑动的大头针的盘子,如果下面有一个物体向上挤压这些大头针,就可以在大头针上端的阵列上复显出该物体的形状。多探针的原子力显微镜也是这样工作的,只不过在更小的尺度上而已。

Feeling Is Seeing

2

WITHOUT A MAGNIFIER, we cannot see objects smaller than a hair. Optical microscopes get us further, but diffraction—the blurring of shadows at sharp edges—kicks in at a few hundred nanometers, and again images blur. We use X-rays and electrons to photograph even nanoscale objects, but we can no longer claim to *see* them: we are semi-blind to the truly miniscule. So: how to navigate the nanoworld?

One approach is the strategy our worm ancestors used—to *feel* rather than to *see*. They lived in mud, where eyes were useless; their skin described their world to them. Feeling is a durable strategy for sensing where there is no light, since it has survived evolution. Cats and catfish use whiskers; cockroaches use antennae.

The atomic force microscope (or AFM)—a finger that brushes surfaces with such delicacy that it can feel individual atoms—was a cornerstone of the foundation of nanoscience. With it, we could *feel* the shapes of molecules! An ecstasy for chemists, whose profession was molecules, but who had never actually seen one.

Of course, the size of a finger limits what it can do: imagine threading a needle while wearing ski gloves. To feel atoms, the tip of the finger must be about the same size as they are. This fingertip looks like a mountain upside down, but it is a shard of silicon sharpened to a point only a few atoms across. To take this photograph, a beam of electrons illuminated the tip.

To draw more complicated pictures by feeling, it is sometimes useful to have many fingers. The child's toy on the following page uses a plate through which many pins slip to replicate the shape of an object pushing up from below. Using many AFM tips at once also works, at a much smaller scale.

3 *Quantum Cascades*

THE WORD "quantum" has a nice sound. It escaped from "quantum mechanics" to become a kind of pet in English. It affectionately attaches itself to unfamiliar nouns—quantum leap, quantum advance—without too much awkwardness. It associates pleasingly with the mysteries of physics. But what does quantum mean? Simply that something occurs only in packages—quanta—having a specific size.

In an ordinary brook, water flows downhill continuously. In this photograph, water flows down the steps of a cascade, with drops in height that are quantized—about a foot for each step. It seems ordinary enough. What startled physicists was to find such steps occurring spontaneously and unexpectedly in nature. Sprinkle salt in the ghostly blue flame of a gas stove, and the flame flares yellow. Look at this yellow light using the right kind of instrument, and you see only a single color: not a mixture of butter and saffron, but exactly one frequency of yellow. Why?

In the early 1900s, no one knew. The astonishing answer was that electrons could not change their energies continuously (like water flowing in a brook), but could have only specific, fixed energies (like water descending this cascade). The difference in energy as the electrons fell from one level to another thus had a specific value (like the difference between the height of water in two levels of the cascade), and this value determined the color of the light they emitted. The discovery that electrons in matter could only have specific energies led to quantum mechanics, turned science on its ear, and ultimately rewrote our understanding of reality.

To appreciate how unexpected quantization was, consider love—something we consider very different from physics. We like to think that love comes in many kinds. "I love you" (because you are my child), "I love you" (because you've just given me a buttered bagel), "I love you" (because in one glance, you completely fried my brains). All are different. But suppose there was only one kind of love, and it came in well-defined packages—quanta of love? Suppose that all the different kinds of love we perceive were just more or fewer packages, or packages of a particular type? We would be astonished, and profoundly disbelieving. Also horrified—as was physics, when quantum mechanics bloomed.

But if it explained the different qualities of love? Even if the story never felt right, we might have to live with the fact that it explained how love works.

量子跃迁

"量子"这个词听起来美妙无比。它现在已经超越了"量子力学"而大红大紫,成为妇孺皆知的一个常用词。人们常常把它加到许许多多的名词之前,像"量子跃迁""量子进展"等等,充满了神秘的物理色彩,但"量子"到底是什么意思?简而言之,就是某样东西以阶梯形式来增减。

在平缓的山溪中,水总是连续地从高往低静静流淌。然而,在右图中,台阶上的水则是跳跃式地向下流动,每次下落的高度是"量子化"的,即每次大概都下降一英尺左右。这似乎是司空见惯的事情。然而,令物理学家们震惊的是,像这样"阶梯式的跳跃"却自发且不可预知地发生在这个世界的每一个角落。例如,将食盐撒在煤气炉的蓝色火焰上,你会看到橙黄色的光,用适当的装置去测量,将发现它并非复合光而是只有一种波长或频率。

为什么只能看到单一波长的光呢? 19世纪初以前,没人可以解释这个现象。最终得到的答案却让我们难以置信:电子的能量不可连续改变,就如同右图中只能在不同高度的阶梯间流动的水一样,电子只能处在分立的、具有特定能量的能级上。当电子从一个能级跃迁到另一个能级,能量的改变是一个固定值,这也正如相邻两个台阶上水的高度差是一定的,而这个能量差决定了发射光的波长或颜色。总之,物质中的电子只能具有特定的能量!这一发现导致了量子力学的诞生,翻开了科学的新篇章,最终改写了历史。

为了理解量子化究竟多么令人难以置信,不妨来考虑一个我们认为和物理完全不搭界的东西——爱。我们知道"爱"有多种原因。"我爱你",因为你是我的孩子;"我爱你",因为你刚刚给了我涂着黄油的面包;"我爱你",因为我对你一见钟情。这些"爱"千差万别。但是如果假设它们都是由一小份一小份相同的、量子化的"爱"所构成的,那会如何?假设我们接受到的"爱"的种类不同只是因为拥有不同份额的量子化的"爱",那又会如何?我们会吃惊、会深深地怀疑、也会恐惧,这正是量子力学刚刚面世时的情景。

但是假如这个量子学说能够解释各种不同的爱,即使对爱的感受已面目全非,我们还得接受这个事实!

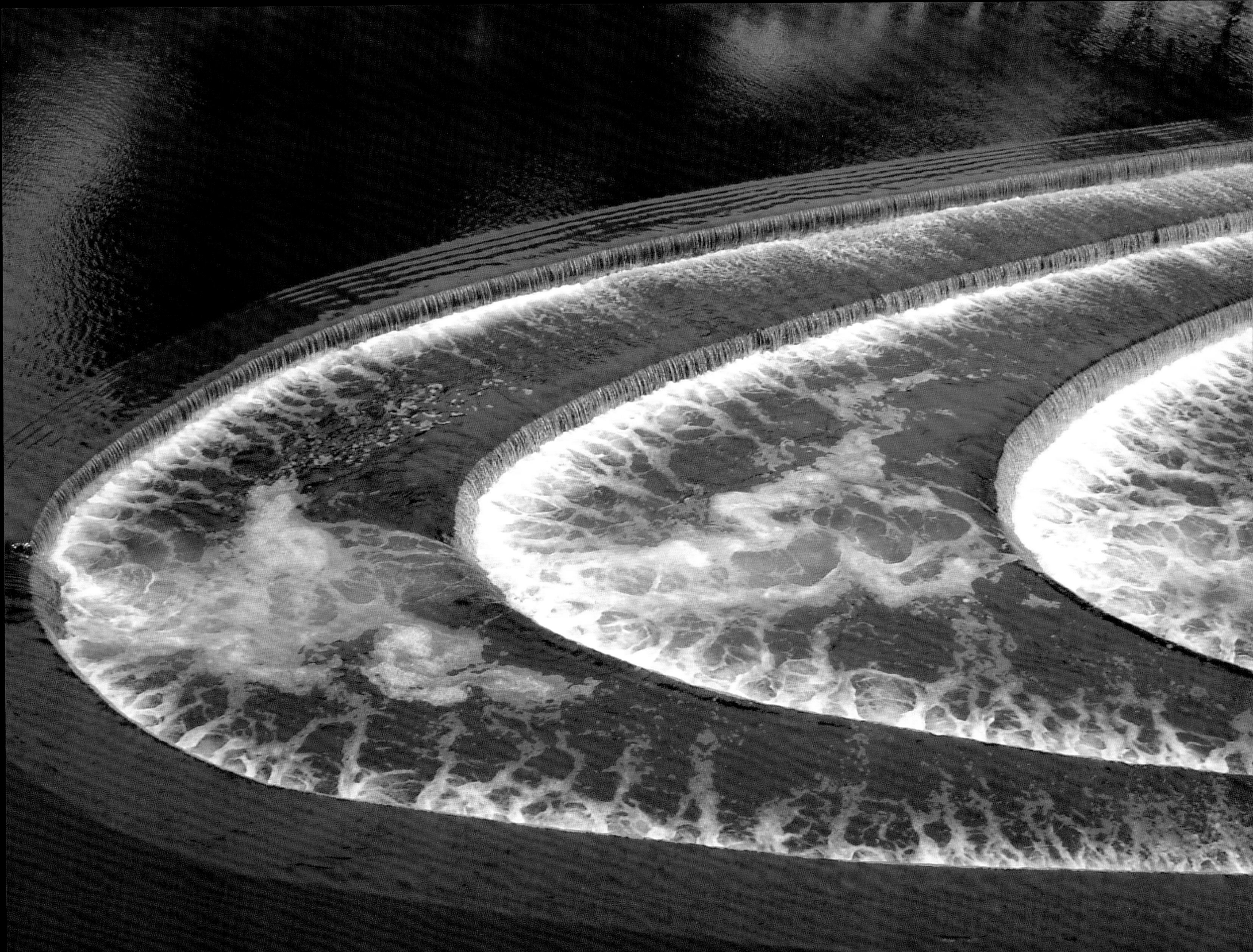

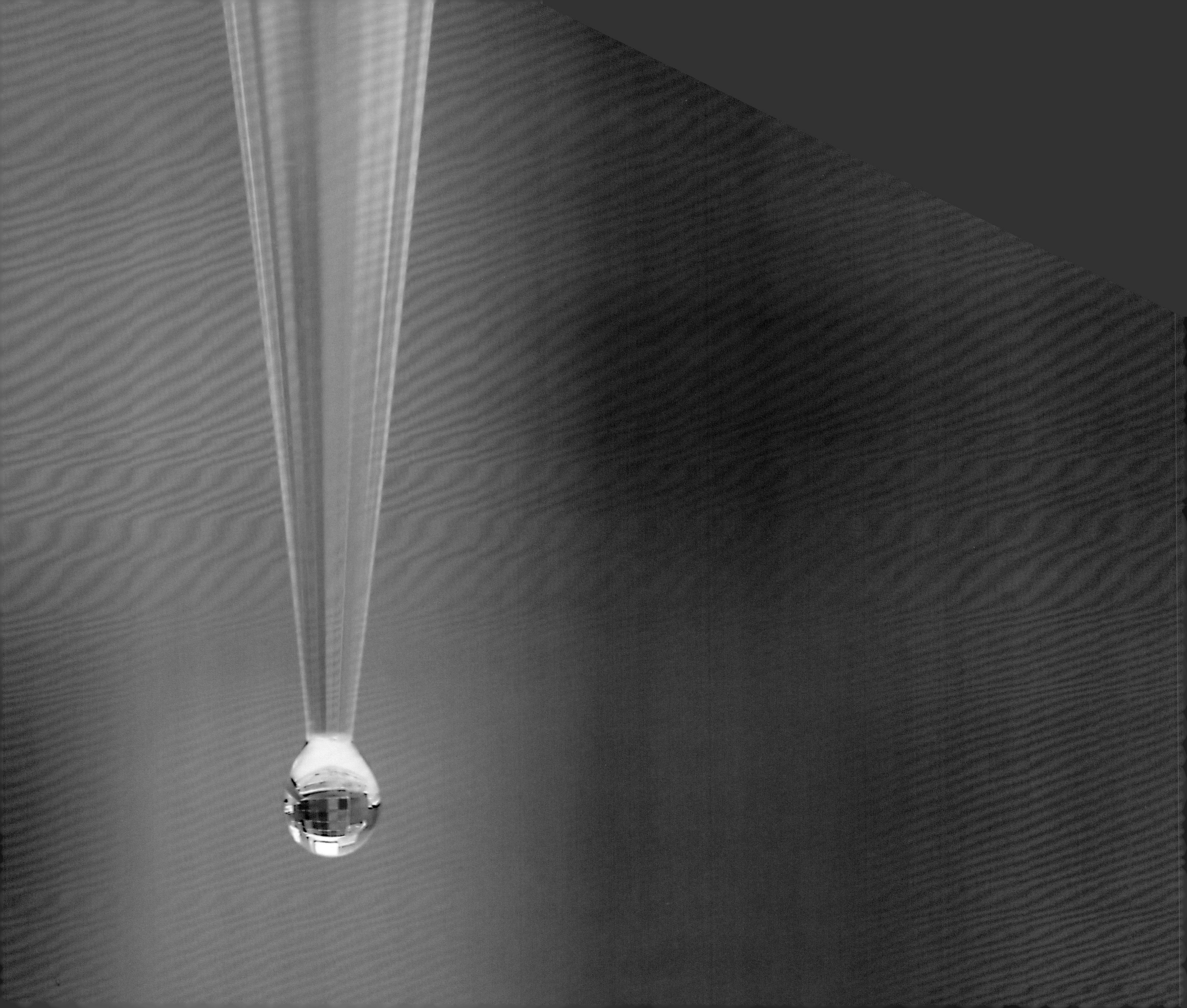

水

海洋以及空气的存在是如此的显而易见，以至于我们常常忘记它们的存在。它们虽然平淡无奇，却不可或缺。倘若没有水、二氧化碳、氧气和氮气这些平淡的小分子，人类将无法生存。我们可以没有金子，但万万不可以没有这些小分子。除此以外，我们还需要阳光来使万物运转、依靠万有引力使万物各居其所。

水分子很小，仅由两个氢原子和一个氧原子构成，可以写成 H—O—H 或者 H_2O。它与其他分子能有很大差别吗？孩子们都有四肢和大脑，都活泼好动，他们之间能有很大差别吗？这两个问题的答案都是"天差地别"。对于化学家来说，水分子如同充满潜力与才能的孩子，虽然很小，但它的两只"手臂"(O—H 键)可以使大量的水分子牢固地结合在一起。它爱憎分明，从不跟油和空气混溶。一般情况下，水分子们都更倾向于呆在水滴的内部，而非边缘或表面。因此只要有可能，水滴总是尽量保持最小的表面积。

为什么吸在移液管口或水龙头上的小水滴不会下落，而牙刷却会？为什么大多数物体都会自动坠落，但小水滴却不会？我们可以从水分子的角度来看这个问题。液滴要坠落必然要向下拉伸，其表面积就会增大，导致更多水分子暴露在表面。而对于水分子来说，这并不是一件"愉快"的事，它们更喜欢缩回球形。这就上演了一场地球引力与液滴表面分子张力之间的拉锯战。

水倾向于降低自身表面积的意义十分重大，它不仅是肥皂泡和细胞膜形成的机制，也是我们在使用沙拉酱前需要使劲摇瓶子的原因。这种性能还使蛋白质能正确地折叠，使分子识别成为可能，这对生命的起源与延续来说起着不可或缺的作用。

那么，除水以外，其他分子能否孕育出生命呢？我们能否用氨气、二氧化碳或者甲烷来替代水，进而构造出新的生物体？这些都无从知晓。谁也说不准在太阳系以外的行星或者这些行星周围的卫星上，是否存在着某些匪夷所思的事情。如果某天在某个遥远的星球上发现散发着立白清洁剂或者苦杏仁味的新生命，我们一定会有所担忧，同时，我们肯定还会很好奇。

Water

4

THE OCEANS, AND THE AIR WE BREATHE, are so *obvious* that we are oblivious to them. But obvious can still be essential. We can't exist without *obvious* molecules: water, carbon dioxide, oxygen, nitrogen. Gold we can do without, but not these. (And, of course, we need sunlight to make it all run, and gravitation to hold everything in place.)

Water is small–just two atoms of hydrogen and one of oxygen: H—O—H or H_2O. How different from other molecules can it be? Well, our children have four limbs, a head, and aggression. How different can *they* be? The answer to both: *very*. To a chemist, water is the little kid with all the talent. It's small, and its two "arms" (the O—H bonds) let it form gangs. It's picky about its friends: water and oil don't mix; neither do water and air. Water molecules much prefer to be insiders, rather than straggle on the edge of the group. Water minimizes its surface whenever possible.

Why doesn't a small drop of water clinging to the end of a pipette, or a faucet, fall? A toothbrush would. Most things fall down, but not a small drop of water. Why not? Think of it from the point of view of the water molecules. For the drop to fall, it must stretch down. But stretching creates more surface, and so forces more of the water molecules to be at the surface. For water, not a comfortable thing to do.

The tendency of water to reduce its surface area is not a minor idiosyncrasy; it is a big thing. It makes soap bubbles and cell membranes what they are, and forces us to shake salad dressing vigorously before using it. It also causes proteins to fold into the correct shape, enables molecular recognition, and is essential to the existence of life.

Might a different molecule–a child with oddities different from those of water–allow a different kind of life? Could life use ammonia, or carbon dioxide, or methane instead? We don't know. As we explore the weird collection of moons scattered around our neighboring planets, and planets circling other suns, we may find stranger things than we can imagine. If we meet a new life form on a distant planet, and it smells like Liby or bitter almonds, we should be wary! And also curious.

Single Molecules 单个分子

5

Since atoms and molecules make up all of physical reality—everything we can touch, taste, see, and feel—it's a little unnerving that we have never actually *seen* one. High school chemistry goes on about atoms and molecules as if they were as ordinary as mosquitoes. In fact, they are *more* ordinary (or at least more numerous), since each mosquito is made up of billions and billions of molecules. Don't misunderstand: we *know* there are atoms, and molecules, and molecules as collections-of-atoms; but a molecule and I have never sat down across a table and looked at one another.

There are, of course, reasons why I cannot see a molecule. First, it is not a solid object that can be seen in the usual sense: it is almost entirely open space, with a few tiny nuclei and a gossamer haze of electrons. Second, I cannot literally see (even with a microscope) anything smaller than the wavelength of light. The shortest wavelength we can see—about 400 nanometers—makes the color violet. The length of a molecule is about 2 nanometers. Much too small. It would be as difficult for our unaided eyes to see a molecule as it would be to see a Great Dane standing on the moon.

So, it's a comfort to have a new way of seeing—if not a molecule in its atomic splendor, at least where one is. Here we "see" molecules—albeit large molecules—using an atomic force microscope. Not perfect, perhaps—we still do not see individual atoms in this image—but certainly better than nothing. These structures are polymers—threadlike molecules perhaps 500 times longer than the molecules of sugar you add to your morning coffee, or the molecules of caffeine that give it its kick. The skins of atoms that make up their surfaces stick to one another in ways that cause one half of the polymer molecule to coil itself into a ball (which appears in this image as a peak), while the other half splays itself across the surface. What we see are the resulting "ball-and-tail" structures: molecular-scale tadpoles.

原子和分子构成了我们所能触摸、品尝、欣赏以及感觉到的一切，可实际上我们从未真正看到过单个原子或分子。这真是令人不安！中学化学就已经开始谈及原子和分子，仿佛它们就像蚊子一样平淡无奇。确实如此，原子和分子比蚊子更为普通，也更为众多。至少，每只蚊子都是由数以亿计的原子和分子组成的。但请注意，我们虽然知道原子和分子的存在，也知道分子是原子的集合体，但分子和我们却从未像家人那样在用餐时面对面坐着，相互凝视。

无法直接看到单个分子的原因有很多。首先，分子并不是一个常规的物体，凭感官就能够看到或感受到。它里面空旷得很，只有一些极小的原子核，被薄雾般的电子云包围着。再者，即便使用显微镜，我们也不能清晰地看到任何比光的波长还小的物体。对肉眼来说，能看到的、波长最短的光大约是波长为400纳米的紫光，而一个分子的尺度大概只有2纳米。在没有特殊装置辅助的情况下，用肉眼观测单个分子就如同用裸眼观测一只远在月球上的大丹狗。

让人欣慰的是，我们已经有了一种新的观察方法。如图所示，我们用原子力显微镜可以观察到一个丝状的大分子。它是一个聚合物，其长度大约是早餐时加到咖啡中的蔗糖分子或者给人提神的咖啡因分子的500倍。这个聚合物中的外围原子相互黏结，使一半的分子蜷缩成球，产生图中显示的那些峰，而另一半分子则沿着固体表面铺展，最终形成了图中“有头有尾”的结构，整个分子如同一个小尺度的蝌蚪。美中不足的是，我们仍旧没有看到分子中的单个原子，但这已经比什么都看不到强多了，至少我们已经能确认单个分子的存在。

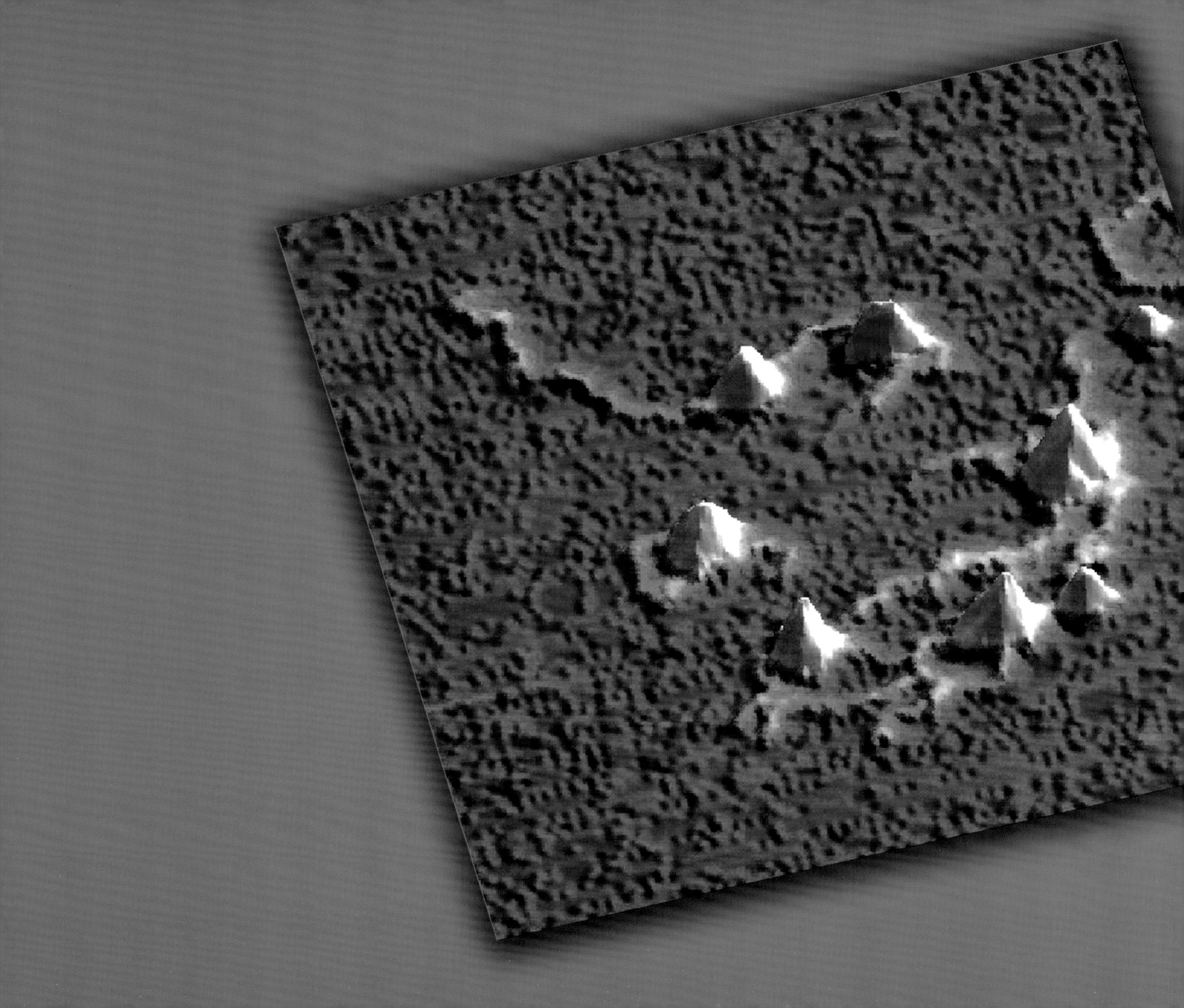

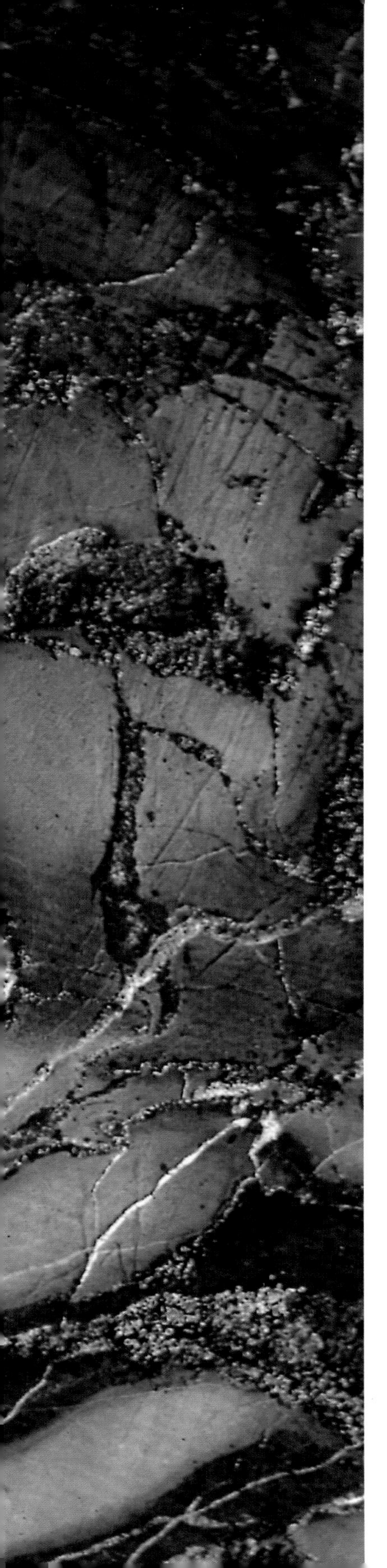

裂　　纹

Cracks

6

地衣是由数个共生微生物组成的菌落，以消化地球形成初期的“骨架”为生。它们生活得十分悠闲惬意，缓缓地从大气中摄取二氧化碳和水分，同时通过溶解岩石获得繁殖所必需的其他元素。随着时间的推移，岩石会慢慢出现裂缝，这与地衣的不断侵蚀有关，有时也会因冰冻时水的膨胀而加剧。对地衣来说，这种冬季常发生的破坏是件好事，这意味着有更多的岩石碎片来供它们消遣。

地衣可没时间去思考裂缝或者断裂是如何产生的。我们也不会，除非有什么贵重的东西破裂了。但是对于断裂，无论是地衣侵蚀花岗岩还是眼镜掉落在浴室的地板上，都是从裂纹开始的。对裂纹如何出现的思考，有助于理解物质在原子尺度的结构。

试想一下，你妈妈留给你的那个唯一的酒杯在杯口出现了一道裂纹。在一端，裂纹大得足以一眼看到，而在另一端，当你迎着光举起酒杯在一个特定的角度观察时，又会发现裂纹消失得无影无踪。在你看到的一端，裂纹的两个表面分离了好几百纳米，而在看不到的一端，裂纹悄悄地融入了杯身，从而造成了消失的假象。在这两端之间又是怎么样呢？

实际上，对于大多数的材料而言，我们对裂纹尖端的纳米结构的认识并不十分清晰。当两个表面相距只有几个原子尺寸时，那些我们熟知的化学原理开始变得不再适用。酒杯上当然不会出现一个比原子还小的裂纹，但物质究竟是如何在原子层次上从分离的两片变成一片的呢？

THESE LICHENS—COLONIES of several symbiotic microorganisms—make their living digesting the bones our planet formed in its infancy. They have a spare life, very slowly extracting carbon dioxide and water from the atmosphere, and dissolving rocks to get the traces of other elements they need to reproduce. As they slowly gnaw away at the stone, they crack it, sometimes aided by the forces of freezing and expansion of water. From the vantage-point of lichens, this wintertime destruction is a good thing—more rock fragments to chew on.

Lichens do not spend much time thinking about cracking or fracture. Nor do we, unless something valuable breaks. But fracture—whether the result of lichens dissolving granite or eyeglasses dropping on the bathroom floor—requires a crack; and thinking about how matter *cracks* leads directly to thinking about its atomic-scale granularity.

Consider a crack in the lip of that last lovely wineglass your mother left you. It can be large enough to see at one end, and disappear in a ghostly reflection—visible only when you hold it at certain angles to the light—at the other. At the visible end, what you see is the separation (on the scale of hundreds of nanometers) of the two opposed surfaces that the crack created. At the disappearing end, the cracks melts into the continuous solid, and there is no separation. What happens in between?

In fact, for most materials, we do not have a very clear idea of the nanoscale structure at the tip of a crack. When the separation between the surfaces becomes only a few atoms wide, familiar chemistry starts to fail. A crack smaller than an atom makes no sense, but how matter progresses from two distinct pieces to one is not obvious at the atomic scale.

Nanotubes

纳 米 管

7

CARBON COMES IN MANY FORMS. One is diamond—hard, cold, colorless, electrically insulating. Another is graphite—soft, slippery, black, electrically conducting. The difference between them is their atomic connectivity. Diamond is a three-dimensional wonder—every carbon atom welded rigidly to four others in endless ranks of repeating tetrahedra. Graphite is like two-dimensional sheets of chicken wire a single carbon atom thick. The three-dimensional strength of diamond is condensed into the two-dimensional strength of graphite. In its plane, a sheet of graphite is probably stronger than diamond, but graphite has no strength at all in the third dimension: one sheet almost does not know that the adjacent one exists. So there you have it: the same atom, different geometries, wildly different properties.

Take a sheet of graphite, roll it up into a tube or a jelly roll, and you have a nanotube—a hollow, electrically conducting cylinder of graphite. To make the geometry clear, a computer generated this image, but the real tube would be a few nanometers across. The center of the tube is hollow: atoms can fit inside.

Nanotubes have properties that leave builders—of bridges, of information processors—dizzy with anticipation: they have strength, rigidity, and electrical conductivity. The last of these is the first to be applied. Electrons slither along the carbon sheets of nanotubes more smoothly than a silk scarf slips through a silver ring. This extraordinary electrical conductivity might be useful for the batteries needed to store solar energy when the sun goes down, or for very small wires to talk to living cells electrically.

碳有七十二变。其中一变是金刚石：坚硬、冰冷、无色、绝缘；二变是石墨：柔软、滑溜、墨黑、导电。这两者之间的本质区别在于它们的原子连接方式。金刚石是一个三维尺度的奇迹——每个碳原子和其他四个相邻原子牢固地连接，构成一个无限延伸的四面体网络。石墨则由一层层蜂窝状的网堆积而成，每层只有一个原子那么厚。从某种意义上讲，金刚石在三维空间的强度在石墨中被压缩到了二维空间上，因此石墨在平面层内的强度甚至高于金刚石，但它在层与层之间的强度却弱之又弱，几乎可以忽略不计。由此可见，同样的原子，由于不同的空间排列结构就会产生完全不同的性质。

如果把石墨片层卷成管状，我们就得到纳米管——一种中空、导电的圆筒形石墨结构。为了更好地描述纳米管的几何结构，我们用电脑画出了这张图片，而实际上纳米管只有几纳米宽。纳米管中间是空的，原子可以被填充在里面。

纳米管无与伦比的强度、硬度以及导电性令工程师们美得不知所措。其导电性是最早被研究的。比起丝巾滑过银戒指，电子“通过”碳纳米管会来得更加通畅。这种优异的导电性能对很多应用十分重要，比方说制备储存太阳能的电池，以及制作给活细胞施加电信号的超细纳米线。

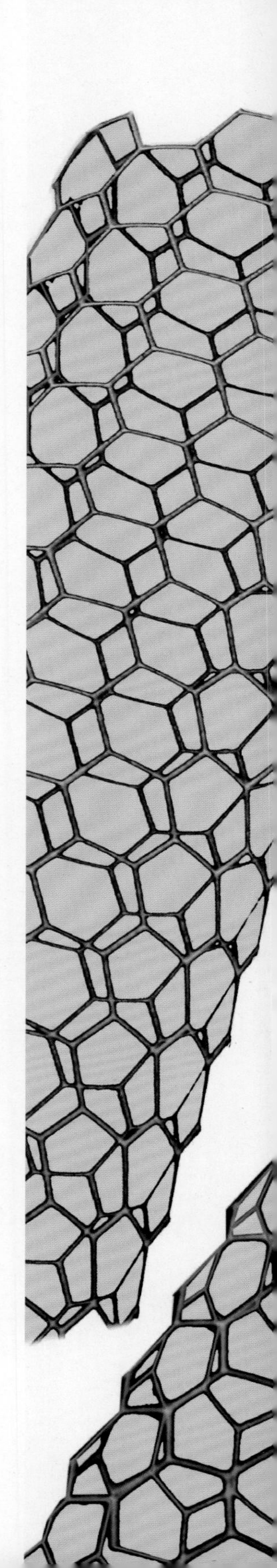

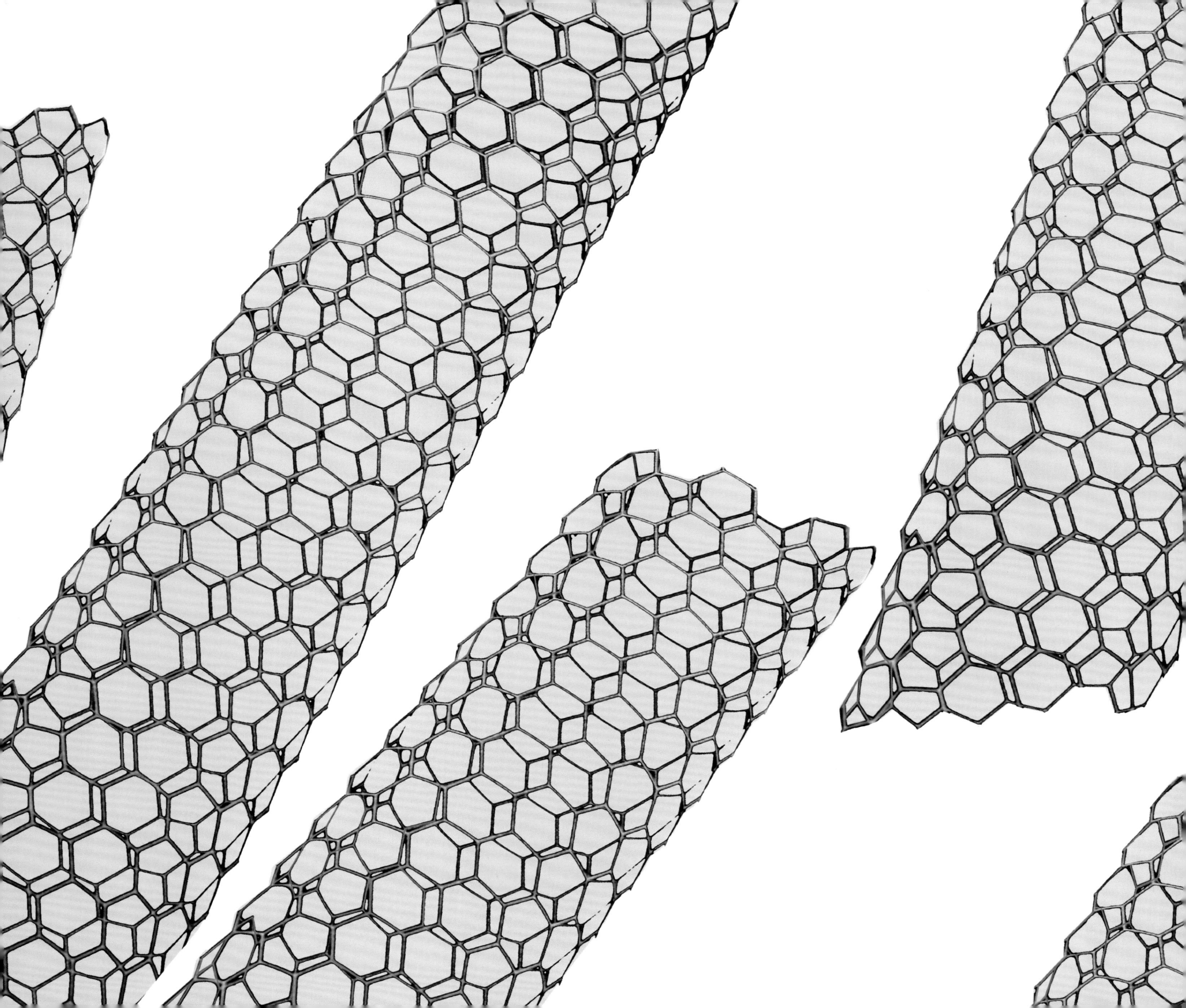

ALICE IN WONDERLAND

What we call "real" is, and then again it isn't.

You—a large object—can be "here" or "there" but not both places at once. A bell, and the sound made by the bell when struck, are different. But our resolute confidence in the ordinariness of experience—in the behavior of things we touch and see—emerges from the extraordinary reality of much smaller things, which are doing something else entirely, and following different rules.

We can use mathematics to rationalize the *pas de deux* of one electron and one proton in exquisite detail. Many electrons and many nuclei we can guess well enough by analogy. But after the mathematics has done its job, we still don't *understand*. Viewed with one eye, an electron resembles a bell; viewed with the other, it resembles the sound of a struck bell.

奇境中的爱丽丝

我们称之为"真实"的东西,不一定就是真实的。

作为一个大的物体,你不在甲地,就在乙地,但你不可能在两地同时出现;一口钟和它发出的声音完全是两码事……这些都是再平常不过的经验。但是这些原本天经地义的东西,却是建筑在不同寻常的小物体基础上的,而后者则遵循着完全不同的规律。

数学能帮我们搞清楚一个电子和一个质子表演"双人芭蕾"的优美细节。通过类比,我们也能对多电子和多质子的体系作出靠谱的推测。但是在数学公式的背后,我们仍然无法理解:为什么从一个角度去看,电子表现得像"钟",而从另一角度,却又表现得像"钟被敲击时所发出的声音"。

Vibrating Viola String

振动的琴弦

8

WHAT DO WE NAME SOMETHING when we don't understand what it is that we are naming?

"Vibrating string" is not simply a "string": a string is an object, a length of cotton or polymer or cat gut. And "vibrating" describes regular back-and-forth motion in anything, from bobbleheads on the dashboard to the thorax of a bumblebee in flight. A vibrating string is a composite of two ideas: "string," an object; and "vibrating," a kind of motion. We have no single word for it, but we understand through experience exactly what it is.

"Electron" is the opposite: we use a single word, but have no intuition for what an electron really is. It seems to be something different depending on how we look. An electron moves like a wave. We understand waves: waves in fields of dry grass in the wind; waves of warm and cool air on a summer evening; waves of perfume as the women pass at the concert. An electron hits like a ping-pong ball (albeit a very small one). We understand balls. But an electron is not just a vibrating ball—it is something else. Sometimes it acts like a wave, sometimes like a ball. Intuition suggests that waves and balls should be different. But intuition is wrong: an electron is *both.*

And if we accept the idea that something can be simultaneously wave and object—if we abandon the notion that our experience with big things should extend, unchanged, to small things, and let mathematics be our counter-intuitive guide—we predict the right outcomes. Still, reality is intensely annoying: our familiar world—the grass, ping-pong balls, the air—is made of molecules, and molecules are made of atoms, and atoms are made of electrons and nuclei. But ping-pong balls behave one way, while electrons and their subatomic friends behave another. How can that be? *Why* should that be?

Physicists like to give pet names to wholly phantasmagorical things: black holes, dark matter, electrons, quarks, neutrons. Pet names for strange pets. Whatever we name them, the smallest bricks of which our substantial world is made have properties that don't sit easily with our experience. That's the way the world *is*; science is slowly getting accustomed to it.

当我们还不理解某样东西时,该怎样给它命名呢?

"振动的弦"并不是简单意义上的"弦"。"弦"是有一定长度的物体,如一段棉线、化纤或肠线;"振动"描述来回往复的机械运动,如仪表指针的摇来晃去,抑或是大黄蜂翅膀的轻盈舞动。而"振动的弦"则同时强调了运动和物体。我们也许无法用一个字来描述这个运动的物体,却可以凭经验想象出琴弦振动的美妙画面。

"电子"则截然相反。虽然它可以简单地用一个字(指英文)来描述,但我们无法想象它是何方神圣。我们看它似山时,它便是山;我们看它似水时,它便是水。电子有时像波一样运动。波是什么?是被微风轻抚的草浪,是夏夜令人躁动的热浪抑或是沁人心脾的凉风,或是在音乐厅中从佳人身上传来的阵阵香水味……一切如此清晰而浪漫。电子有时也像乒乓球一样弹来弹去,虽然它比乒乓球小了许多。球还是比较好理解的。然而,电子并不仅仅是一个振动的小球,它是一个很难说得准的东西。它时而像列波,时而像个粒子。虽然直觉告诉我们波和粒子是风马牛不相及的东西,但直觉也有错的时候,对电子而言,它既是波,也是粒子。

如果我们接受物质的波粒二象性,如果我们抛弃利用宏观认识来描述微观体系,如果我们背弃直觉而选择让数学推演来引导我们,那便可正确地预测结果。尽管如此,现实依然让我们无法接受:我们所熟悉的草地、乒乓球、空气等这个世界上所有的一切都由分子组成,分子又由原子组成,原子则由电子和原子核组成。那为什么电子和它的亚原子朋友们能够体现出波粒二象性,而乒乓球却仅仅表现出粒子性?这怎么可能?这究竟又是为什么呢?

物理学家喜欢把那些变幻无常的事物亲昵地称为黑洞、暗物质、电子、夸克、中子等。这就像是在给一些奇怪的宠物命名。无论怎么称呼它们,这些构成我们现实世界的最小"砖块"天生就被赋予了不太容易和我们的经验相符的性质。世界就是这个样子的,科学正在逐渐揭开它的庐山真面目。

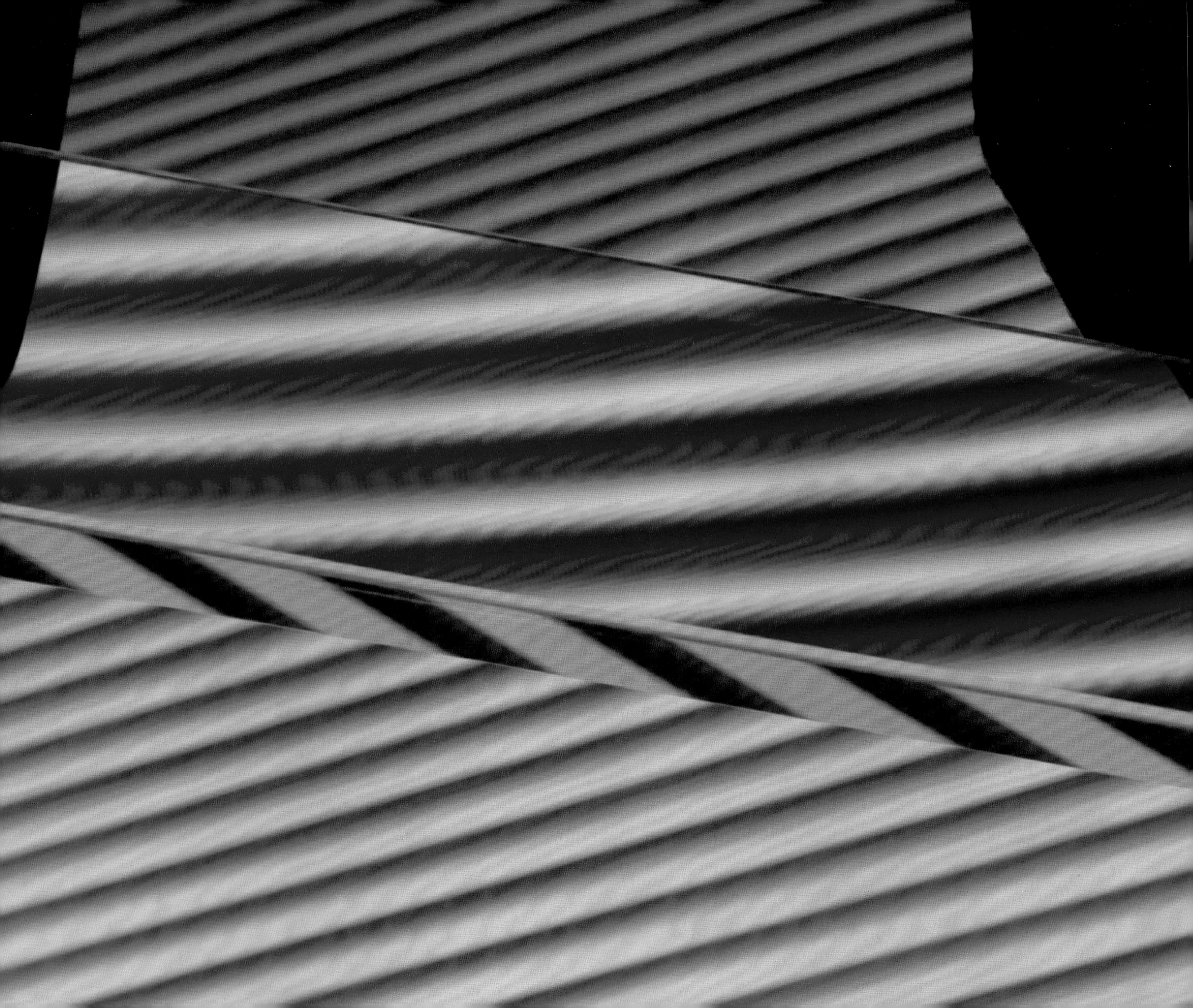

棱镜和折射

Prism and Refraction

9

白光通过棱镜折射所产生的图案就像孔雀开屏般绚丽多彩，赤橙黄绿青蓝紫，令人目不暇接。其中紫光弯折的程度最大，红光则最小。这种利用折射将白光分散成不同颜色确实让人着迷。还有比这更有趣的吗？

不过，什么是光？就像空气、水、重力一样，这些太熟悉的东西，我们往往会轻易忘掉。退一步讲，光也许什么都是——微型猎枪弹、氤氲的雾气、连续的波；也许什么都不是，光就是光，它源自电荷的加速运动。

对棱镜而言，光的行为像是波。为了理解光与棱镜之间的相互作用，也就是折射，你必须先了解一下棱镜。其大体的物理图像还是比较简单的：对于光来说，棱镜像是一组带电荷的弹簧振子（电子与原子核的吸引和排斥就像是弹簧）。轻轻地推动电子，它便开始振荡；另一方面，由于电荷的加速运动，导致了电磁场的变化，反过来又产生了光，但其相位和传播方向与原来的光相比会略有差别。这就如同随波逐流的圆木随着水波上下浮动，但不会与水波的节奏完全吻合。将光试想为波，这些波在棱镜中穿越时会诱导电子的振动，而振动的电子又会产生许多更弱的波。如此类推，你便能很容易地理解，为何当光从空气进入棱镜（入射）和从棱镜再度进入空气（出射）时，都会被弯折，且因为红光比蓝光的弯折度要小，所以白光可以被分成五颜六色。这就有点像海洋的波浪沿着防浪堤被弯曲。

光的行为十分古怪。通过棱镜或者眼睛的晶状体时，它就像波；当打在视网膜上时，它又成了一颗子弹。简单来说，它运动时是波，与其他东西相撞时则是粒子。

A SLAB OF WHITE LIGHT becomes a spectral peacock's tail—red, yellow, green, blue, violet—as it bends through a prism. Violet bends the most; red the least. The resolution of white—the mixture of all colors—into separate colors by bending is gaudy. Is it more interesting than that?

Well, what *is* light? Air, water, light, gravity—all parts of reality so familiar to us that we forget to remember them. Light *might* have been anything. Tiny BBs of something; a fog of something else; ripples in a continuous ether of yet another thing. It isn't any of those: it's *light*, which isn't like anything else we know. It's what happens when electrical charge accelerates.

As far as a prism is concerned, light behaves like waves. To understand the details of the conversation between the light and the prism—of the diffraction of light—you have to understand prisms. The general picture is simple enough. To light, a prism is a set of charges—electrons—on springs. (The attractions and repulsions among the electrons and nuclei act like springs.) Light nudges the springs and makes the electrons oscillate; that movement in turn generates light—charges are accelerating—but slightly out of phase. It's like a floating log bobbing with the waves, but not *exactly* with the waves. Imagine light as waves, and imagine that these waves make the electrons in the glass of the prism bob as they pass, and that the bobbing electrons make more, small waves. Add it all up, and you can convince yourself that light should seem to bend as it crosses from air to prism, and bend again in going from prism to air; red light should also bend less than blue. It's a bit like an ocean wave bending around a breakwater, but only a bit.

And there's the odd behavior of light. Going through a prism, or the lens of your eye, it behaves like a wave. Going into your retina, it behaves like a BB. It's a wave when it moves; it's a particle when it hits.

10 Duality

波粒二象性

WE'RE BURDENED BY A curious conditioning that blinds us to one of the greatest—perhaps *the* greatest—of art forms. We live for poetry; we live in terror of equations.

We see a poem, and we try it on for size: we read a line or two; we roll it around in our mind; we see how it fits and tastes and sounds. We may not like it, and let it drop, but we enjoy the encounter and look forward to the next. We see an equation, and it is as if we'd glimpsed a tarantula in the baby's crib. We panic.

An equation can be a thing of such beauty and subtlety that only a poem can equal it. As an evocation of reality—as the shortest of descriptions, but describing worlds—it is hard to beat the most artful of poems and, equally, of equations. They are the best our species can do.

Equations are the poetry that we use to describe the behavior of electrons and atoms, just as we use poems to describe ourselves. Equations may be all we have: sometimes words fail, since words best describe what we have experienced, and behaviors at the smallest scale are forever beyond our direct experience.

Consider Margaret Atwood:

You fit into me
Like a hook into an eye
A fish hook
An open eye

Consider Louis de Broglie (a twentieth-century physicist, and an architect of quantum mechanics):

$\lambda = h/mv$

Read the equation as if it were poetry—a condensed description of a reality we can only see from the corner of our eye. The "equals" sign is the equivalent of "is," and makes the equation a sentence: "*A moving object is a wave.*" Huh? What did you just say? How can that be?

It's an idea worth trying on for size. Poetry describes humanity with a human voice; equations describe a reality beyond the reach of words. Playing a fugue, and tasting fresh summer tomatoes, and writing poetry, and falling in love all ultimately devolve into molecules and electrons, but we cannot yet (and perhaps, ever) trace that path from one end (from molecules) to the other (us). Not with poetry, not with equations. But each guides us part way.

Of course, not all equations are things of beauty: some are porcupines, some are plumber's helpers, and some *are* tarantulas.

我们往往对诗情有独钟,而对方程式充满畏惧,这让我们对"方程式"——这一可能最伟大的艺术形式视而不见。

看到一首诗时,试着品味它:读上一两行,在脑中反刍,看看它是否合我们的胃口。假如不喜欢的话,不要紧,忘掉就行,我们仍会回味那刹那间的邂逅,并对下一首充满期待。而当看到一个方程式时,我们就像突然见到婴儿床上有一只凶相毕露的毒蜘蛛,感到惊慌失措。

然而,有一个方程式它是如此的精妙绝伦,只有诗歌才能与之媲美。它用最简短的语言描述了世界的真谛,即使是最优美的诗歌也难超越它。这真是一个人间奇迹!

就像用诗来描述我们自己一样,我们用方程式来阐述电子与原子的行为。语言可以形象地描述我们经历过的事,但在小尺度的世界里,面对那些远远超出我们认知经验的行为,语言会突然显得异常的苍白无力,只有方程式才能表明事实的真相。

玛格丽特·阿特伍德在诗中写道:

你融入我
如钩入眼
一枚鱼钩
睁大的眼·

而德布罗意(20世纪的物理学家,量子力学的奠基人之一)则写道

$$\lambda = \frac{h}{mv}$$

如果我们像读诗一样来读这个方程,含义将一目了然:"="的意思是"是",左边是波,右边是一个运动的物体,连起来便成了"一个运动的物体就是一列波"。哦!你在痴人说梦吗?这怎么可能?

这是一个值得思考的问题。人性尚可使用语言通过诗歌来描述,而方程式所描述的事实则超出了语言的表述范畴。演奏赋格曲,品尝新鲜的西红柿,写上一首诗,甚至谈一场恋爱,这一切的一切都源自于可用方程式描述的分子和电子。但我们却还不能在人性和事实之间找到关联,不管是通过诗,还是通过方程式。它们都从某种程度上引导着我们的生活。

当然,不是所有的方程式都如诗如画。有些像豪猪,令人望而却步;有些像通下水道的工具,单调乏味;还有些则像塔兰图拉毒蜘蛛,长相可怖却色彩绚烂。

$$\lambda = \frac{h}{mv}$$

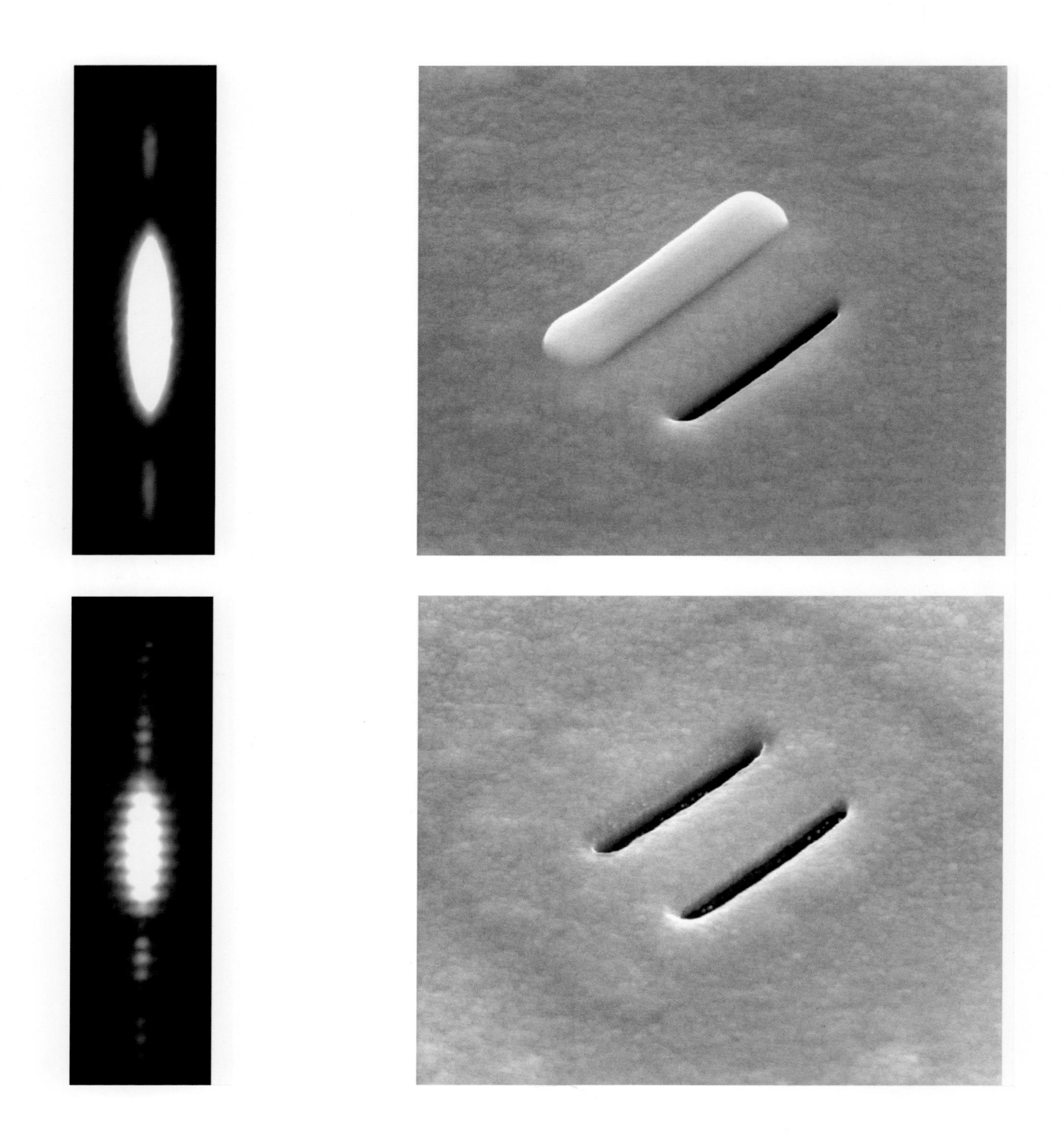

波的干涉

Interference

11

试想这样一幅美景：阳光，沙滩，海浪。而你和着浪花的节奏，迈着轻盈的脚步。清凉的海水没过脚踝，安抚你躁动的心绪。但浪涛如世事无常，当浪头约会浪尾，浪花只能轻吻你的脚趾；当浪头邂逅浪头，浪花则会拥抱你的小腿。在你惬意地享受假日时光的同时，大自然也正在讲道说法——波的干涉（相消与相长）。

科学研究就是把事物拆得七零八落，然后再把它组装回去。就好像把一栋房子拆成钢筋、水泥和砖块，再重建一栋更漂亮的。各个科学分支都在研究大千世界的某一部分：物理学用数学抽象地描述原子核、电子以及原子；化学以原子为“砖块”，构建分子和材料，甚至细胞；生物学从细胞入手研究生物体；社会学以人为基础研究社会现象……每个学科都有自己的一套。

我是一个化学家，我的世界就是原子核和电子，以及它们的无数种组合。原子处于常规的宏观世界和不可思议的量子世界之间。其中，电子最令人费解，有质量和电荷，却没有大小。它不像原子那样金屋藏娇（指原子核），在它的里面空空如也。读到这里，你也许会问：“如果什么都没有，那么它的质量和电荷又从何而来呢？”问得好！

如图所示，电子可以表现得像波一样。图中光滑的椭圆是大量电子通过单个纳米尺度的狭缝后在探测器上形成的图案。而有涟漪的椭圆是大量电子通过两条平行的，相距几纳米的狭缝后形成的图案，这个现象跟波通过相同狭缝时所出现的干涉完全吻合。虽然量子物体的波看起来与海浪大相径庭，但实际上它们有着很多的相似之处。

作为一个化学家，我对电子和光子的许多怪异性质，以及它们如何掺和在一起构成宏观物体，常常感到百思不得其解。有些时候，当我凌晨四点在陌生的旅店醒来，低声问自己究竟懂得多少时，答案常常是：几乎为零。

IMAGINE A BEACH: SUN, WAVES, SAND. Wading in the wash of the waves along the beach—cold water up to the ankles—calms frayed nerves. Waves are unpredictable. If the crest of one wave falls into the trough of another, just a few riffles slither up over your toes; if the crest of a big one falls on the crest of another, it's water up to the knees. That's part of the fun. It's also an example of *interference*—destructive and constructive—between waves.

Science likes to take the world apart and then put it back together. Start with a house; tear it down to boards and nails; use them to build a better house. Each field of science takes one part of the problem. Physics starts with mathematical abstractions and builds to nuclei, electrons, and atoms. Chemistry starts with atoms, builds molecules, and continues to materials and cells. Biology builds organisms out of cells. Sociology builds societies out of organisms. Each field does its thing.

I'm a chemist. My universe is nuclei and electrons, and the almost endless ways they can assemble. Atoms are just at the border between ordinary, macroscopic matter and matter dominated by the Alice-in-Wonderland rules of quantum mechanics. Electrons, in particular, have the unnerving property of having mass and charge but no *extent*—no size. There's no tiny BB down in their core, as there is a nucleus sitting at the center of an atom. "Ah," you say, "that's strange. If there's nothing there, what is it that has a mass? And what's charged?" Good questions.

These images show that electrons—whatever they are—can behave as waves. The smooth oval is an image formed on a detector by electrons going through a single nanoscale slit. The oval with ripples is formed by electrons going through two parallel slits separated by a few nanometers: it's what you expect for interference between waves passing through two channels in a nanoscopic breakwater—the chop in the waves in a quantum harbor. The wave behaviors—including interference—of quantum objects is different from that of ocean waves, but there are also similarities.

As a chemist, I've come to uneasy terms with the weirdnesses of electrons and photons, and with their ability to meld into the ordinariness of macroscopic things. But sometimes, lying awake in a strange hotel room at 4 a.m., considering what I might say that I *really* understand about anything, I fret that the answer is: almost nothing.

Quantum Apple 量子苹果

12

HERE'S AN APPLE. It's an interesting, odd, glass apple, but no matter: it's at least the shape of an apple. It casts a shadow.

Fine; but if you look again, you will see that something's wrong. The shadow is not what we expect from an object having the shape of an apple; in fact, it isn't what we expect from anything. Part of it seems to be the shadow from an apple, part from a cube. So what is it? Is there a square cross-section that we can't see? The answer is that the image has been altered; the apple and its two shadows are irreconcilably at odds.

The image is cheating: it gives the appearance of reality but is synthetic—a metaphor for a quantum object, which also has attributes that seem irreconcilably at odds. In our world, a grain of rice is a completely different kind of thing from the ripples on a pond. With quantum objects—electrons, photons, atoms, and molecules—these nice distinctions are not so useful. A thing can be simultaneously a particle and a wave (or it can behave—when looked at in different ways—as a particle or as a wave). In fact, a particle is *necessarily* also a wave. Even you and I are waves, albeit waves with crest-to-crest distances of 10^{-35} cm (a distance so small that it is unimaginable). A quantum apple could perhaps also simultaneously be a quantum cube.

We live in a world of large things. A 10-pound concrete block weighs 10 pounds and is made of concrete, regardless of how we look at it. Small things are more elusive; what you glimpse depends on how and when you look. Even the act of looking changes the thing that is looked at.

图中所示是一个苹果,一个有趣的玻璃苹果。虽然看起来有点诡异,但起码是苹果的样子,旁边还有它投射的影子。

如果再仔细点,你会看到一些不对劲的地方。它的影子并不像我们所期待的那样像一个苹果的形状,而是部分像苹果,部分像一个立方块的投影。这一切似乎在告诉我们,玻璃苹果可能不像我们看到的那样真的是个"苹果",那它到底又是什么呢?难道它有一个我们看不见的正方形截面吗?答案是这幅图案已经被人为地修改过了,这个苹果和它的影子是相互矛盾的。

这张图片极具欺骗性,它虽然反映了部分事实,但却是人工合成的。我们不妨用它来比喻一个量子物体,因为量子物体本身就是个矛盾综合体。在我们的世界里,一颗米粒和池塘里的涟漪是完全不同的东西,但是对于量子物体,如电子、光子、原子、分子等,宏观世界里那些显著的区别就不再适用了。一个物体可以同时既是粒子又是波,或者说当我们用不同的方式去观察它时,它可以表现得要么像粒子要么像波。事实上,所有粒子都必然同时是波,甚至你和我都是一种波,只是波长小到了超出我们想象的程度而已,仅有 10^{-35} 厘米。如此一来便不难理解,一个量子苹果也许同时也是一个量子立方体。

我们生活在一个宏观的世界里。一个十磅重的混凝土砖块,不管我们怎么测量,它都是十磅重并且是由混凝土制成的。但是微观物体则难以捉摸,你看见的取决于你怎么去看和你何时去看,甚至连看一下也会改变你看的东西。

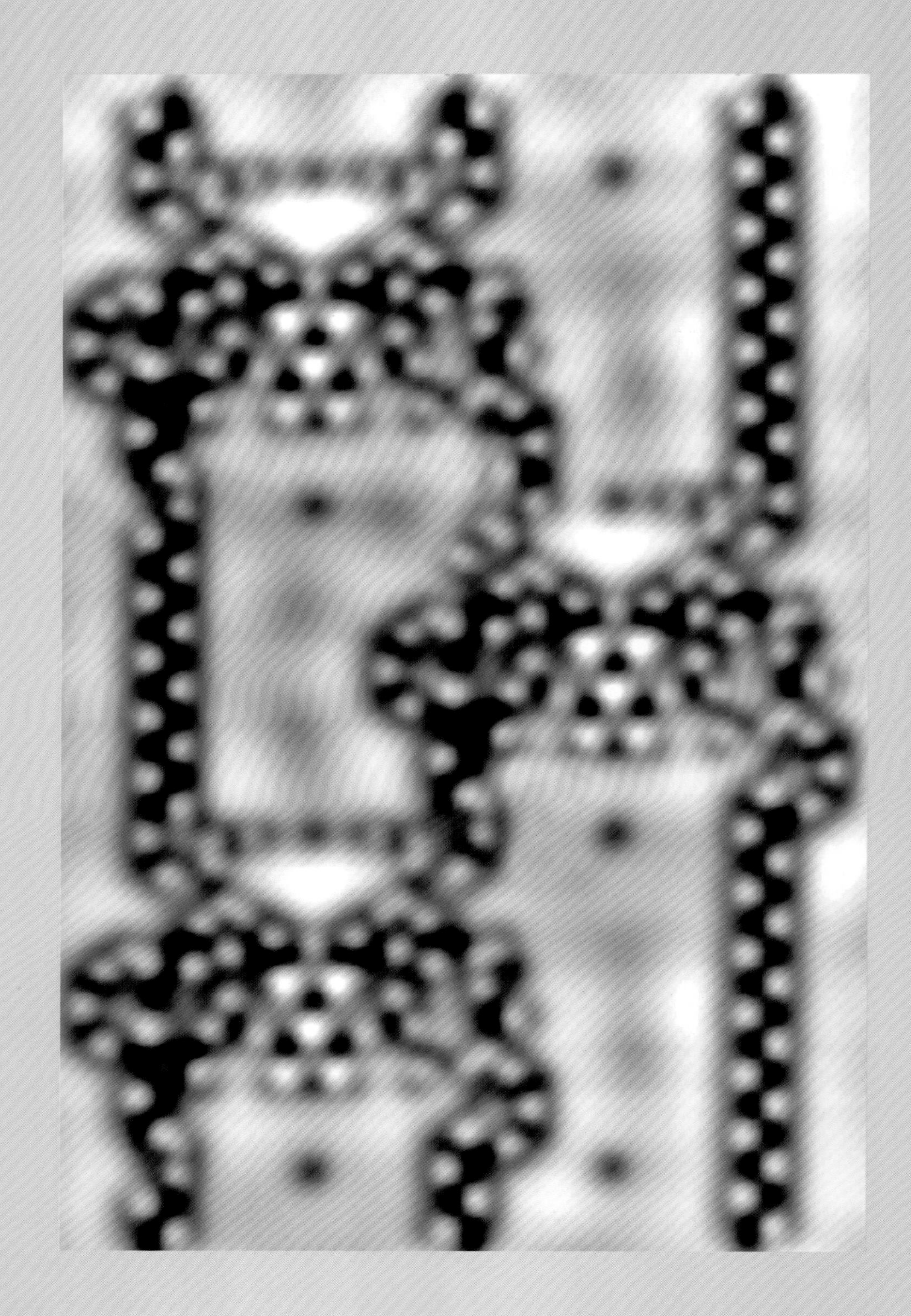

分子多米诺

Molecular Dominoes

13

开关无处不在且至关重要，但我们很少会留意它们。当我随手扳动墙上的开关，便能在巨大的电网系统中打开一个极小的新通路，随即点亮床头灯。交通信号灯是十字路口处控制车流的开关；手机语音信箱里的留言提醒我给汽车加油，这一简单的过程却需要操控数十亿个开关来重新配置信号通路，好让我的手机在浩瀚的电磁波中找到属于自己的点滴信息；吸一小口咖啡，我的神经网络就会在舌尖和大脑间相互交谈并传递信息，而作为开关的突触将神经细胞一个接一个地联接起来，从而让肌肉可以对我的想法作出相应的反应……

所有能控制输出的东西都是开关。电灯开关控制电流，交通信号灯控制车流，淋浴器的手柄控制水流。在某些情况下，开关甚至也可以是一本打开的支票簿或是一个试探性的吻。

开关可以小到什么程度呢？它们的工作原理又会多么稀奇古怪呢？这张图片展示的是一氧化碳分子，相互紧靠着，在金属薄膜的表面勉强保持着平衡，只要推倒其中一个，其余的就会一个接一个地跟着跌倒。最终形成的结构则取决于它们在表面的分布图案以及推动的方式，整个过程就如同推倒一组精细布置的分子多米诺骨牌。分子们居然能表现得像孩子们手中的多米诺骨牌玩具一样，真是难以置信！

到目前为止，用亚稳态的分子组合作为开关还只是个游戏，还不能用来解决任何实际问题，就连这种想法也像训练跳蚤跳古典芭蕾一样匪夷所思。然而，蛐蛐都可以被训练成角斗士，谁又能断定跳蚤就学不会芭蕾呢？

SWITCHING IS EVERYTHING. When I flick the wall switch, I open an infinitesimal new pathway in the vast circulatory system of the continental power grid, and turn on my bedside light; I give the switch not a second thought. Traffic lights switch the flow of cars at intersections. The message on my voicemail reminding me to put gas in the car requires billions of switches to reconfigure the circuits that route the flows of bits, and let *my* cell phone find *my* trickle of bits in the electromagnetic smog in which I unwittingly marinate. I take a sip of coffee: marvelous networks of nerves in tongue and brain chatter to one another, and connections between them—synaptic switches connecting one nerve cell to another—make my muscles serve my whim.

Anything that controls an output is a switch: a light switch controls electricity, a traffic light controls traffic, and the handle on the shower controls water. All are switches. So, in some circumstances, are an open checkbook or a tentative kiss.

How small can a switch be? How strange can its operation be? This image shows *individual* molecules of carbon monoxide, precariously balanced in patterns on the surface of a metal film: push one, and the next falls over, and then the next. The end result is a cascade that depends on both the pattern and the push: it's like the collapse of cleverly configured molecular dominoes. Astonishing! Molecules behaving like children's toys.

Using metastable arrangements of molecules as a switch is, so far, more a game than a practical solution to a problem. But the *idea*—to arrange sets of molecules to serve as a mechanical switch—is as quirkily wondrous as training fleas to perform classical ballet.

14 The Cell in Silhouette

细胞肖像图

THE CELL IS THE SMALLEST LIVING THING. What quarks are to physics, atoms to chemistry, and bits to computer science, cells are to biology. They are the stuff organisms are built of. They are as simple as anything in biology; they are as complicated as anything we know.

Much of science starts with the briefest of sketches—the equivalent of a silhouette of Abraham Lincoln, or the Classic Comic version of *War and Peace*—and then fills in details. Here is a kind of silhouette of a mammalian cell. The round part was its nucleus—the container for DNA. The nucleus stores the cell's job description, its instruction set for replication, its memories of its ancestors. The rest was a balloon—the cell membrane—filled with the myriad molecules that give the cell its personality and do its work. The wispy tendrils suggest internal structures. As a silhouette, it's not bad: sack, nucleus, internal machinery.

But this shape—a square—is purest artifact. What biologists study—the cell in a Petri dish—usually has a form that has little to do with the shape a cell would have in an organism. A square is particularly unnatural: this cell has been glued into place using clever chemistry. The cell is probably no more comfortable stretched "square" on a Petri dish than was a heretic stretched "long" on the rack, and its behavior may be no more natural. The relevance of the answers it gives to questions about the normal behavior of cells in organisms is also suspect.

细胞是最小的生命单位。它对于生命科学的意义,就如同夸克之于物理学,原子之于化学,比特之于计算机科学。细胞构成了生命体,它跟生物学里面的许多东西一样简单,但又比我们所有熟悉的事物更为复杂。

科学通常是从最简要的草图开始的,就像是一幅亚伯拉罕·林肯的轮廓图,或者是《战争与和平》的经典漫画版,都是先将大致的框架勾画出来,然后再丰富其中的细节。图中显示的是一种哺乳动物细胞的轮廓图,圆形的部分是脱氧核糖核酸(DNA)所在的细胞核,用来储存细胞的功能信息、复制指令以及遗传物质;细胞膜则像个膨胀的气球,里面充满了许多具有独特性和功能性的分子;而图片中的那些须状物则对应细胞的某些内部结构。这幅图简单明了地勾画出了细胞的神态:包括细胞膜、细胞核以及某些内部结构。

不过,对于生物学家用培养皿长出来的细胞而言,它的形状通常与器官里面的细胞截然不同。这个正方形的细胞表现得尤为突出,它是使用了一种巧妙的化学方法将其困在特定的图形中而形成的。无论是把细胞扯成方的或拉成长的,都不会让细胞感到舒适,而细胞的行为也会跟体内的正常细胞大相径庭。如此说来,用这种体外培养或者外界操控的办法来研究细胞行为,是否能真实地反映出细胞在生物体内的活动规律,还需要进一步考证。

Laminar Flow

15

When two rivers meet, they mix. Eddies form; the swirls carry tendrils of water from one into the other, the tendrils diffuse together, and the rivers lose their separate identities. For two streams of water in microchannels, the rules are different. When they meet, there are no eddies, and they do *not* mix (or do so very slowly).

The flow of fluids in microchannels is *laminar*—that is, the fluid flows as if it were a sheet or slab rather than a liquid. Small channels, low rates of flow, and high viscosities favor laminar flow. The colors in this image are streams of water, flowing in channels approximately 100 microns wide—the width of a hair. The direction of flow is from top right to bottom left.

Arms of a glacier show the same behavior when they meet; so do the slow flows of honey. Even people in narrow corridors move in an orderly stream; being close to a wall keeps everyone in line. If the corridor is wide, people meet, stop, chat, and swirl. Controlling flows in microchannels is vital to microfluidic systems used in biotechnology to analyze blood and urine, to grow cells, and to crystallize proteins that are targets for new pharmaceuticals.

层　流

当两条河相遇，它们交汇时形成的涡流会带着卷须状的水从一条河流入另一条，进而迅速地扩散开去，于是河流便失去了它们原本的特征。对于在微通道中流动的两股水流，它们所遵循的规律却与河流大不一样。当它们相遇时，不会有涡流产生，也不会混合，或者说混合的速度十分缓慢。

微通道中液体的流动是层状的，也就是说此时的液体好似板块而非流体。在小的通道内，低的流动速度和高黏度的流体更倾向于形成层流。这张图片中，五颜六色的液体在大约100微米宽（约为一根头发的宽度）的微通道中流过，它们正从通道的右上端流向左下端。

冰川的支流相汇时显示的行为与微通道中的流体相似，缓慢蠕动的蜂蜜亦是如此。甚至在狭窄通道中移动的人们都可以形成有序的人流，因为通道太窄以致于人们不得不一个接一个地紧贴着墙移动。如果通道很宽阔，人们就会邂逅、逗留、交谈，甚至成群结队地通过。有效地控制微通道中液体的流动行为对生物技术至关重要，例如血液和尿液分析、细胞生长，以及将可能成为新药的蛋白质结晶，等等。

16 The Wet Fantastic

湿的奇谈

WHEN TWO CARS CRASH head-on, shards scatter in every direction. The forces of collision tear metal and shear connections; once a piece is loose, nothing pulls it back into place. It's a chaotic event.

If two streams of *water* collide (albeit with less energy), the results are different and fantastic. In this image, two thin jets of water run into one another almost, but not quite, headlong; (in the photograph, one stream almost hides the other). They deform under impact by spraying laterally into a smooth sheet. The sheet spits out fingers and drops of liquid at its edges, and its core then contracts back into a central stream. The complexity and symmetry of the pattern were completely unexpected.

The difference between cars and these streams of water is partially the energy of collision, partially the way water responds to pressure and gravity, and partially surface tension. When streams of water collide, the liquid responds to the pressure in the region of collision by squirting in the only directions it can: laterally. Gravity also pulls it down. Surface tension—minimization of surface area—makes the sheet contract into fingers and then into drops, and finally collects the remnants of the sheet again into a continuous stream of droplets.

We often associate complex behaviors—the spontaneous formation of intricate patterns, unexpected changes over time—with systems that are themselves complicated. Two streams of water are as simple as one can get, but still—see what happens! Even the simplest systems have the potential to show behaviors that confound us.

当两辆汽车对撞时,碎片会散得满地都是。即使是金属部件,在巨大的撞击力面前也只能毫无反抗地被撕裂或变形。哪怕一个小小的零件松动了,也无法自动恢复原状。撞车就是如此混乱无章。

但如果是两柱水相撞,结果就奇妙和有趣得多了。如图所示,两股细流奔向对方,它们不是像粗暴的公牛那样迎头相对,倒像是久别重逢的情人,相互投入对方的怀抱(在图像中,一股水流几乎被另一股遮挡了)。这两股相涌的水流形成一层纵向的界面,在界面周围散发着零星水滴,落向两旁,而中间的部分则缩回水流。整个过程如焰火般璀璨夺目。对这种碰撞带来的如此复杂而又对称的图案,我们只能叹为观止。

汽车和水流的碰撞结果如此不同,既要归因于其相去甚远的撞击力,也有水对压强及重力的响应因素在内,当然还有表面张力的作用。当两柱水对撞时,受碰撞区域压力的影响,水从纵向流动变成横向流动,产生一层薄薄的水膜,重力则迫使其坠落。同时水流的表面张力会尽可能减少其表面积,从而使一部分水膜变为小水滴而散落,其余部分则成为一串相连的水珠。

我们通常把复杂的现象归结于体系本身的复杂性,例如随时间变化莫测的复杂图案。而两柱水流是再简单不过的体系,但我们仍然能从它们的碰撞中观察到十分复杂的现象。由此可见,再简单的体系也可以让我们费解。

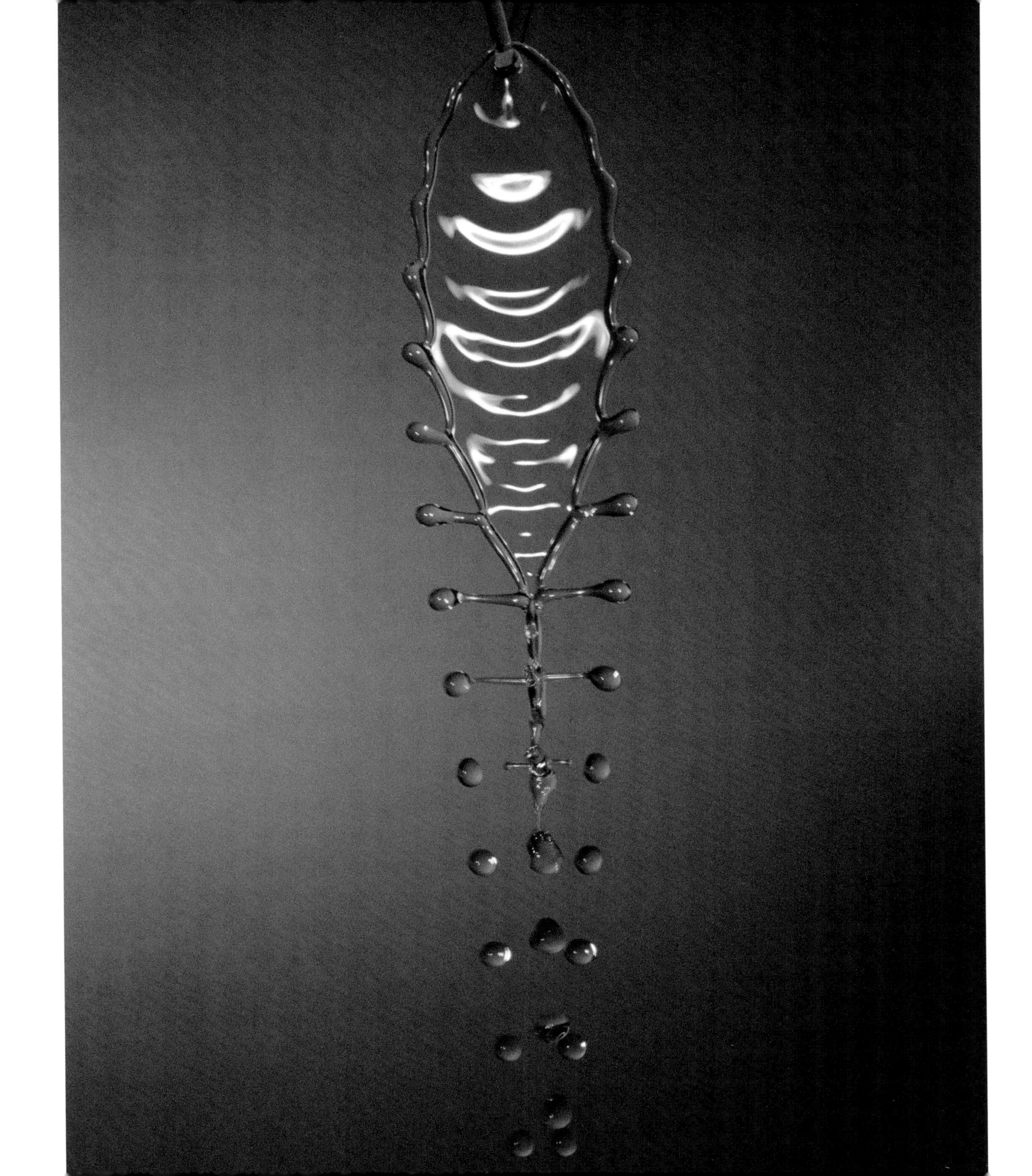

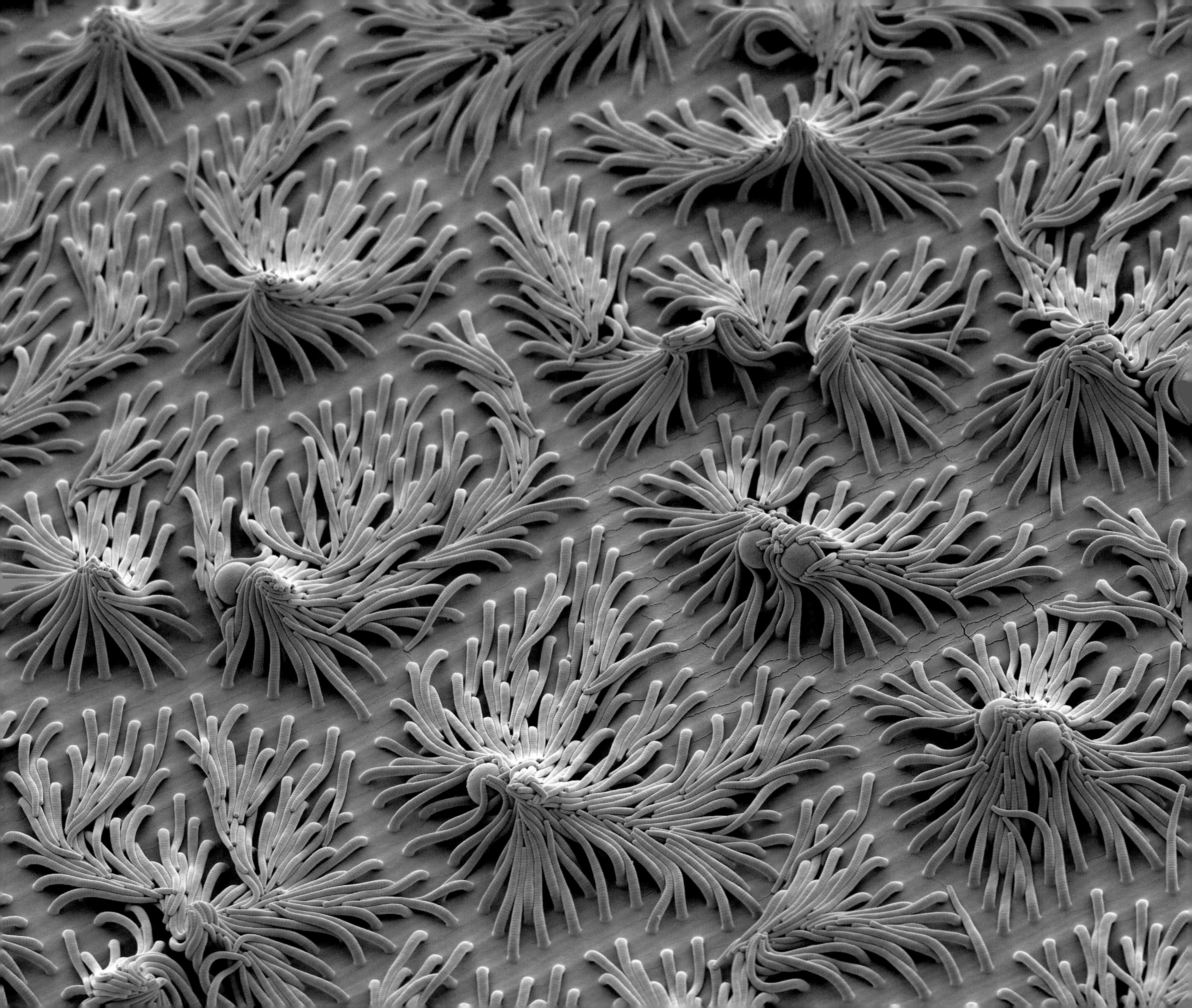

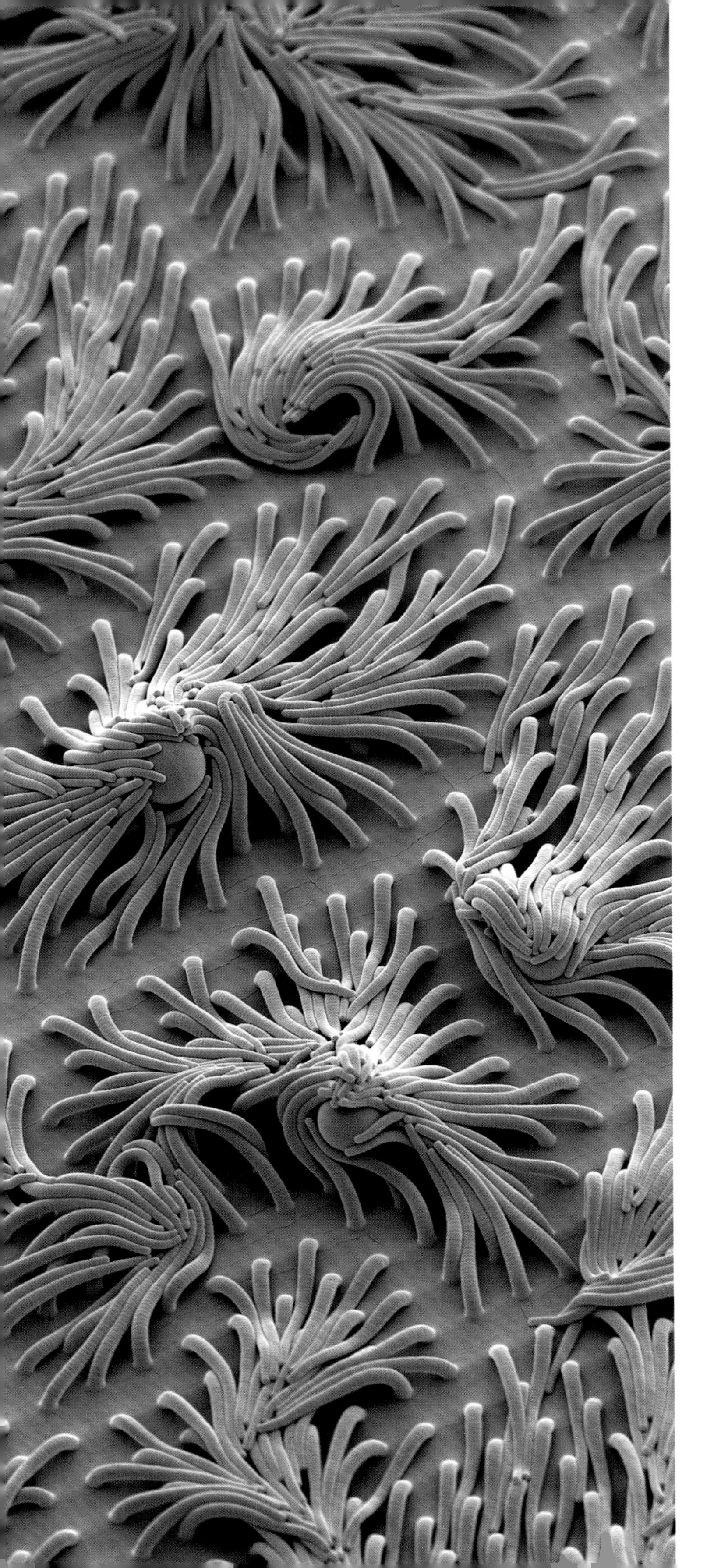

Fingers

17

MOTHERS SOFTLY SURROUND THEIR children with their arms. Their children grip their favorite ball with all the fierce strength their little fingers can summon. Both are protecting what they love.

These fingers are fronds of polymer—wisps a thousandth the width of a hair. The balls around which they so gracefully drape are small polymer spheres. The spheres originally floated as a suspension in water. A drop of this suspension was placed on the surface sprouting the fronds. As the water evaporated, the spheres settled on the fingers, and capillarity—the tendency for the last, thin film of water to form beads rather than to spread—pulled the fingers to the balls.

It's easy to be confused by inanimate things when they mimic behaviors with which we have emotional associations. Anthropomorphizing capillarity into affection or avarice is misleading but unavoidably appealing.

"手　　指"

妈妈会温柔地把孩子拥在怀里,而孩子则会使出吃奶的力气紧抓住他(她)最喜爱的皮球。保护自己的所爱正是人的天性。

图中的这些"手指",虽然它们细如发丝,也能牢牢抓住自己的"所爱"——聚合物微球。起初,这些微球是悬浮在水中的,而这些手指则被固定在一块基底上,像刚从土里探出头来的嫩芽般欣欣向荣。取一滴含有微球的水放在同一基底上,随着水的蒸发,微球逐渐沉积到基底表面。在最后的阶段,水会缩成众多的小液滴,而水滴的表面张力牵引着这些"手指"伸向微球,并将其紧紧抓住。最终的结果让人目瞪口呆,这些"手指"看起来好像原本就附着在微球上似的,无论你怎么努力,也休想让它们分开。

看到这些"手指"睿智的表现,你不免要怀疑它们是否真的拥有生命。虽然把这些无生命的物体人格化可能会让我们在追寻真理的过程中误入歧途,但我们却往往会情不自禁。

LIFE

We are each a flame.

A stove burns wood in air. Its heat makes breakfast—fat spitting in the frying pan, water boiling in the kettle.

We burn the sugars we eat (sugars not very different from wood), using the air we breathe. Admittedly, the flame we are heats a million intricately connected pots and kettles, and the meal they make is us: growth, reproduction, aging, personality. We *are* more complicated than breakfast.

And the smoke that curls and twines from our many fires is what we call "society."

生命

我们每个人都是一团燃烧着的火焰。

当木柴在空气中燃烧，它提供的热量让水在壶中沸腾，把咸肉中的脂肪炸出来，最终给我们一顿早餐。

我们则利用空气来燃烧体内的糖分。从某种意义上来讲，糖和木柴并没有太大区别。所不同的是，用我们的"生命之火"和复杂的"锅碗瓢盆"做成的大餐却是：成长、繁殖、老化以及人格，这可比做好一顿美味的早餐要复杂得多！

而所谓的"社会"不过是我们这些"火焰"冒出来的"烟"罢了。

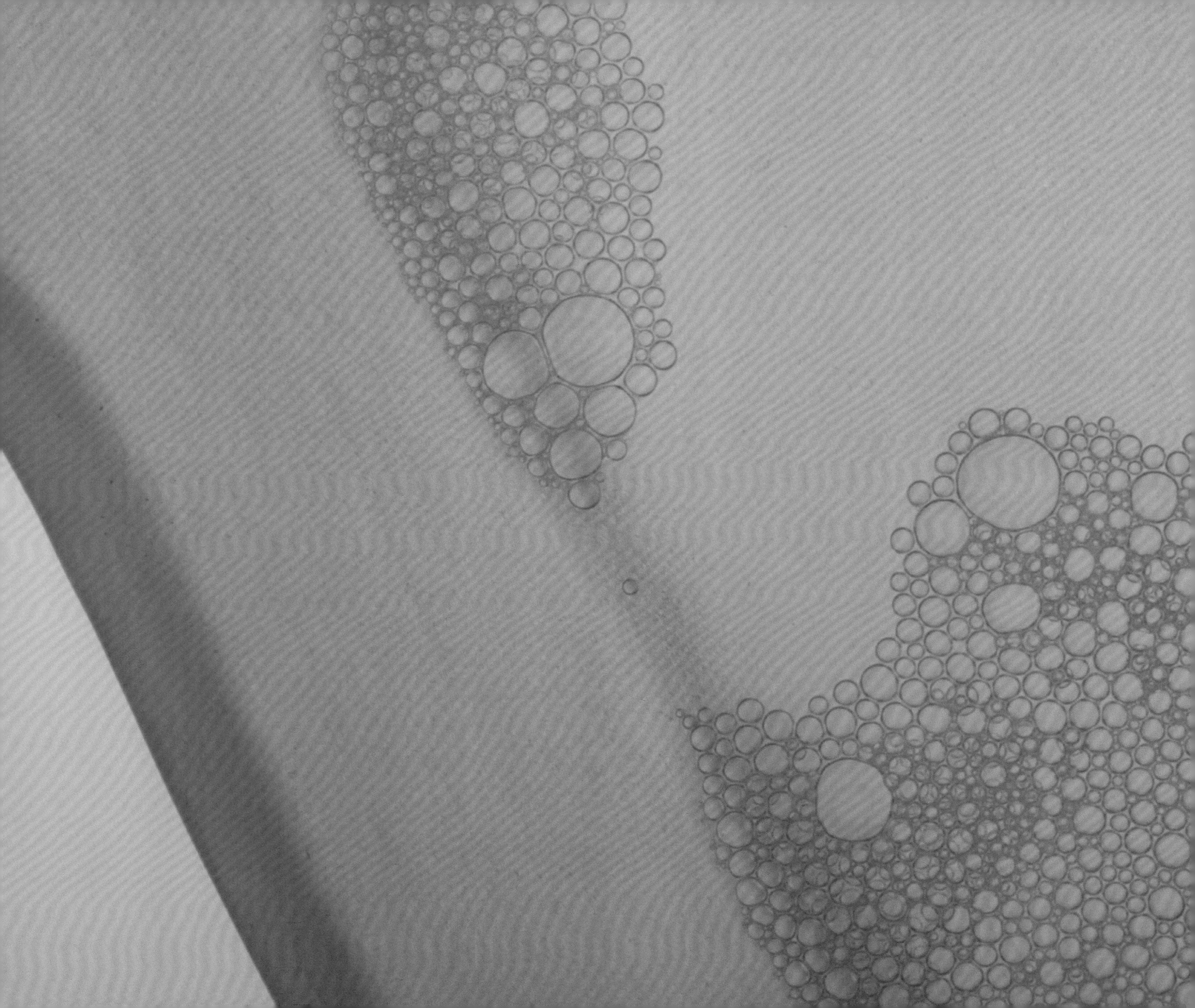

肥　皂　泡

Soap Bubbles

18

细胞表面的膜与肥皂泡的膜差不多，只是内外表面相反而已。它们构造的原理都是基于水和油的不相溶性。

想象这样一幅画面：肥皂分子就像两端性质截然不同的牙签一样，一端亲水，而另一端亲油。细胞膜中的分子亦是如此。当你把这样的一个分子投入水中，其中一端会迫不及待地与水结合，而另一端却总是想逃逸。那么这个分子该如何调解这样的矛盾呢？不必担心，它们会很快自发地形成单分子层，将亲油的一端集合在一起并远离水分子，同时亲水的一端与水相接触。如此一来，这个矛盾综合体便表现得爱憎分明了，爱的一面相濡以沫，恨的则老死不相往来。

我们可以用抹了黄油的面包片来演示这个过程。面包片本身是亲水的，而涂上黄油的一面则亲油。当我们把两张面包片中抹了黄油的那一面贴在一起，这样做成的三明治就好像细胞膜的横截面，即内部有一层油性的、十分薄的膜（对于细胞膜来说，只有两个分子层那么厚），而面包部分则在外面并被水包围着。如果把这种片状的三明治变成一个球状物，你就得到了与每个细胞表面膜相类似的结构。这个连续的油层把细胞的里外分开，里边是生物分子的水溶液，外边则是充彻了细胞间隙的流体，其主要成分也是水。

肥皂泡的情况恰好相反。把两片面包抹上黄油的面都露在外面与空气接触，再在这两片面包之间添上一层水，肥皂泡便大功告成了。这层水有几千个分子层那么厚，比细胞膜的双分子层要厚得多。当然，这个厚度相对宏观尺度来说还是挺薄的。

尽管细胞膜和肥皂泡的基本结构都体现了油水互不相溶的事实，但它们之间还是有天壤之别的。最主要的一点的就是肥皂泡仅仅由水和肥皂分子组成，而细胞膜中则镶嵌了其他的功能性分子。这些分子中，有的就像时开时合的嘴巴一样按一定频率控制着其他分子的进出；有的却像针线一般连接着相邻的细胞或者是细胞与它们所处的环境；还有的则监测着细胞外部环境的变化，并及时地将信息汇集到细胞内部。

THE FILM THAT DEFINES THE SURFACE of a cell is similar to the film that forms a soap bubble, but turned inside out. Both structures are based on the fact that oil (or fat) and water don't mix.

To get the picture, imagine a molecule of soap as a toothpick whose two ends have incompatible properties: one end is attracted to water, and one to fat. (The molecules in a cell membrane are similar.) When you bring such molecules into contact with water, half of each molecule is *attracted* to the water, the other is *repelled*. How do they solve this dilemma? Remarkably, they start by spontaneously forming molecularly thin sheets, with the fat-loving ends all together on one side and the water-loving ends on the other.

Imagine the result as a piece of bread and butter: the bread (the water-loving part) on one side, the butter (the fat part) on the other. Now, in your mind, bring two bread-and-butter slabs together butter-to-butter: this sandwich suggests a cross-section of the membrane of the cell: on the inside, a very thin film of fat—about two molecular layers thick—with bread on the outside, surrounded by water. Extend this sandwich into a balloon, and you've got the sack that surrounds every cell. The continuous film of fat is what separates the inside of the cell (a solution of biological molecules in water) from the outside (intercellular fluid, which is also mostly water).

For a soap bubble, do the opposite. Put the two layers of bread together with the butter on the outside, in contact with air, and then add a film of water between the pieces of bread. The water is thicker than the bread-and-butter layer—a few thousand molecules thick—but to us, it is still very thin indeed.

Although the basic structures of cell membranes and soap bubbles show two aspects of the same phenomenon—separation of oil and water—the differences are vitally important. The biggest is that a soap bubble is just a film of water and soap, while the cell membrane is studded with other molecules. Some of those open and close like mouths to regulate the furious molecular traffic entering and exiting the cell; some stitch the cell to its neighbors, or to the molecular jelly in which it sits; others monitor what is happening on the outside, and send reports home to the inside.

The Cell as Circus

细胞与马戏团

19

IT IS A NICE SIMPLIFICATION to think of the cell as a balloon filled with salty, watery soup, with some strands of spaghetti called DNA floating in it, and some beans called proteins. As I said, a simplification.

This image shows a cell from the top, splayed out across a flat surface. It has been killed, embalmed, and dyed to make various parts of its anatomy show up. What appears is a little like a photograph of a circus tent. The green strands are molecular ropes: they attach it to the surface, prop it open, and hold it together.

But if it were a circus, we wouldn't see the animals, the clowns, the ringmaster in tails, the beautiful lady in pink-and-sequins-and-feathers on the elephant! And certainly not the egos, quarrels, and intrigues. It is impossible to get a sense for the vitality of the circus—of the roles and idiosyncrasies of the animals and the performers—from a satellite picture of an abandoned tent. It is impossible to appreciate the elegance and feverish activity of the cell from its image after it has been pickled and rouged.

There are many ways other than microscopy to understand cells. DNA sequencing opens a kind of sketchy instruction book. Analysis of proteins says who the workers are, but only hints at what they do, and says nothing about how they work together. We can't tell the bosses from the animals. The ticket takers operate using an arithmetic we do not understand.

It's all quite mysterious. But then we are all still rubes at this circus, and dazzled by all the action.

细胞可以简化成一个气球，里面充满了盐和水分，并漂浮着一些通心粉似的DNA分子，还有一些像豆子一样的蛋白质。当然，这只是个十分简化的模型。

这幅图显示的是一只已经凋亡并被舒展开来染上颜色的细胞。它的形状宛如马戏团的帐篷，绿色的线条就像分子绳索，牢牢地将帐篷系在地上，并将帐篷撑开，保持整体稳定。

在这样的马戏团里，我们看不到动物、小丑、穿着燕尾服的马戏团领班以及骑在大象上、身着华丽衣服的漂亮女郎，更看不到人们之间的钩心斗角和自私自利。当然，我们不可能从一张废弃帐篷的卫星照片中，感受到马戏团曾经的辉煌。同样地，我们也不可能从凋亡细胞的图像中，感受到它昔日从容的生命和勃勃的生机。

除了显微镜，我们也可以用其他方法来研究细胞。DNA测序是我们开始了解细胞的指令书。通过对蛋白质进行分析，我们可以知道它们是谁，大概是如何工作的，但至于它们之间是如何相互作用的，我们还毫无头绪。当然，光看到这些蛋白质，还不能搞清楚谁是真正的主宰者。另外，细胞用的"定量"方式至今还让我们像丈二的和尚摸不着头脑。

细胞的世界里迷雾重重。在这里，我们就仿佛刘姥姥进了大观园一样，对细胞马戏团里绚丽的表演只有瞠目结舌的份儿。

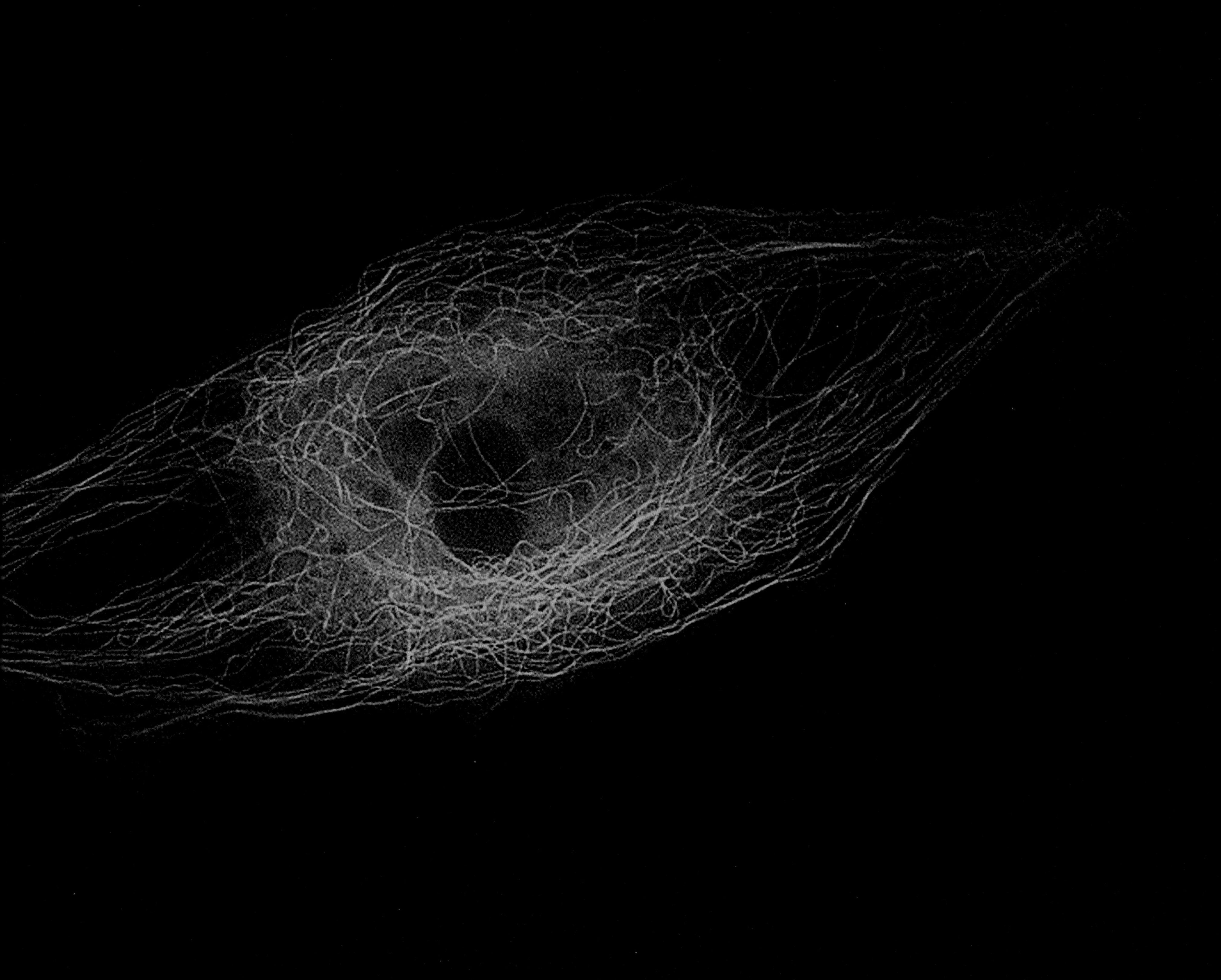

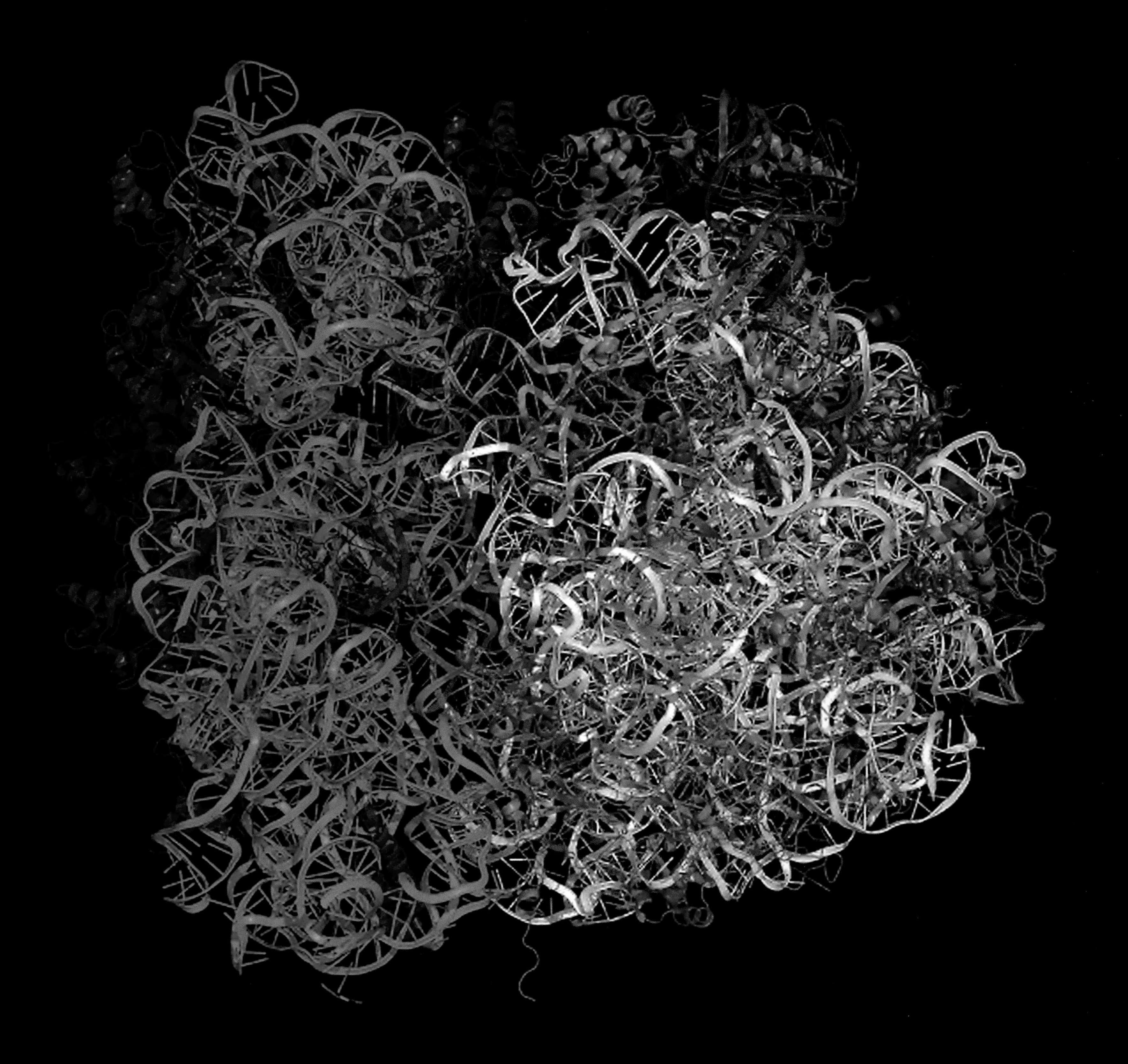

核 糖 体

Ribosome

20

与工业流水线上组装汽车的机器人相比，细胞中的很多分子能在更为精密的纳米尺度下完成更为复杂的工作。核糖体（图中只展示了核糖体的一部分）是我们所知的最大的功能大分子之一，它是由蛋白质与 RNA 组成，直径约 20 纳米。核糖体的结构异常复杂，它毕竟是经过上亿年的优胜劣汰进化而成的。

细胞里布满了组织严谨但又杂乱无章的分子群。在这里，核糖体把一种一次性的“磁带”——从 DNA 中转录而得的信使 RNA——嵌入到自己内部并用来指导合成一种全新的分子，即蛋白质。20 种不同类型的氨基酸分子是构成蛋白质的基础，按照信使 RNA 指令，核糖体以每秒数以百计的速度把特定的氨基酸依次串在一起，合成出蛋白质分子。

核糖体合成蛋白质是在水溶液中进行的，这就比生产汽车的机器人更为灵活。另外，在合成过程中也不用担心重力的影响，也不需要传送带。令人匪夷所思的是，合成蛋白质也不需要电脑来控制，但却是一种全自动化的生产模式，它所用的分子机理目前人类还只能望洋兴叹。我们不禁会问：自然界是怎样打造出种类和机制如此复杂但又十分精密的分子结构的呢？

目前还没有人知道答案。核糖体的进化过程如童话故事般跌宕起伏：尝试、纠错、再尝试、再纠错……在如此反反复复的几亿年间，它在不断地自我完善和更新。达尔文为我们勾画出了这个进化蓝图，但是等我们明白过来的时候，故事已经发展到了第 100 章，我们无法回头从第 1 章读起。时至今日，我们还没有完全弄懂达尔文的理论，其中许多的细节还有待认识，更不用说完全理解了。

THE CELL CONTAINS MANY molecules that carry out nanometer-scale operations more complex than the meter-scale jobs robots do on an automobile assembly line. The ribosome (of which this image shows only a part) is one. It is among the largest functional molecular structures we know—an aggregate of protein and RNA perhaps 20 nanometers in diameter. This molecular hairball is impossibly complicated—the accretion of parts and patches from billions of years of Darwinian optimization.

The belly of the cell is a scene of organized molecular chaos. Here, the ribosome finds and inserts into itself a disposable molecular tape (messenger RNA, transcribed from DNA), and uses its instructions to synthesize an entirely new sort of molecule (a protein), by adding twenty component pieces (molecules called amino acids), one at a time, in sequences dictated by the RNA, to a growing string. All this at rates of hundreds of additions per second.

The ribosome assembling proteins has an advantage over a robot assembling cars—the parts it uses float in the water that surrounds it: it does not contend with gravity or conveyor belts. It also has no controlling computer: it somehow operates autonomously, according to molecular principles we can vaguely appreciate but cannot even faintly mimic. The obvious question is: How did evolution produce a structure of this enormous complexity and sophistication?

The answer is: We don't know. The evolution of the ribosome is a tale of trial and error that has been developing for several billion years. Darwin sketched the storyline for us, but we are reading Chapter 100 without being able to read Chapter 1. As with any great story, Mr. Darwin's excellent idea has layers of subtlety that we have not yet even recognized, much less understood.

Bacterial Flagella

21

WE UNDERSTAND INTUITIVELY how submarines move. There's a hull, a motor, a shaft, and a propeller. Straightforward. The motor turns the shaft; the shaft turns the propeller; the propeller pushes against the water; the submarine goes forward. We can simulate the forces on the propeller blade by moving a hand rapidly through the water in a swimming pool.

Our intuitions about the behavior of water on a large scale fail for water on a small scale. To a bacterium, water is like cold honey; a turning propeller moves it nowhere. So it uses a strategy very different from that used to move a submarine. The bacterium has many motors, and each turns a long, flexible, helical whip called a flagellum. When all the motors turn in the same direction, the whips wind themselves into an orderly helical bundle: as it spins, it drills the bacterium forward. This bundle is more like an auger than a propeller.

A bacterium has no steering motor and no rudder. It changes direction by turning several of its motors into reverse. The helical bundle unwinds, and the bacterium spins randomly. It then puts all of its motors into "forward" again, and sets off in whatever direction it happens to be pointing.

It sounds hopelessly inefficient; but there are uncountably more bacteria on earth than there are humans. They are small and many; we are large and few. In the end, they eat us. Who is more efficient? Who wins?

细菌的鞭毛

我们对潜水艇的驱动原理并不陌生。潜水艇主要由船体、引擎、杆轴和螺旋桨构成。它的工作原理也简单易懂：引擎带动杆轴转动，杆轴则带动螺旋桨旋转并向后排水，从而使潜艇前进。在游泳池里，我们快速划动手臂，也可以得到类似于螺旋桨产生的前行动力。

我们不能把对水的宏观认识直接推广到微观世界。对于细菌来说，水就好比冻结的蜂蜜一样黏稠，仅靠一个推进器根本无法在其中动弹。因此，细菌采取了与潜水艇完全不同的策略。细菌长有很多波状弯曲的鞭毛，细长而灵活，每一根鞭毛都是一台"引擎"。当所有的引擎都朝同一个方向转动时，这些鞭毛就聚成一束有序的螺旋体。当它整体旋转时，细菌便向前移动。比起潜水艇的螺旋桨，这一束螺旋状的鞭毛更像是把螺丝刀。

细菌没有方向盘和舵，它是通过反向调节一些引擎来改变其前行方向的。当螺旋状的鞭毛束不转时，细菌便随波逐流；当它缓过劲来时，又会启动所有引擎，向前进发。

细菌的运动方式听上去效率很低，但是别忘了，细菌的数量是远远超过我们人类的。它们虽小，却数量庞大；人个子很大，但数量却有限。最终，我们还是逃脱不了被它们吞噬的命运。谁的效率更高？谁是最后的赢家？

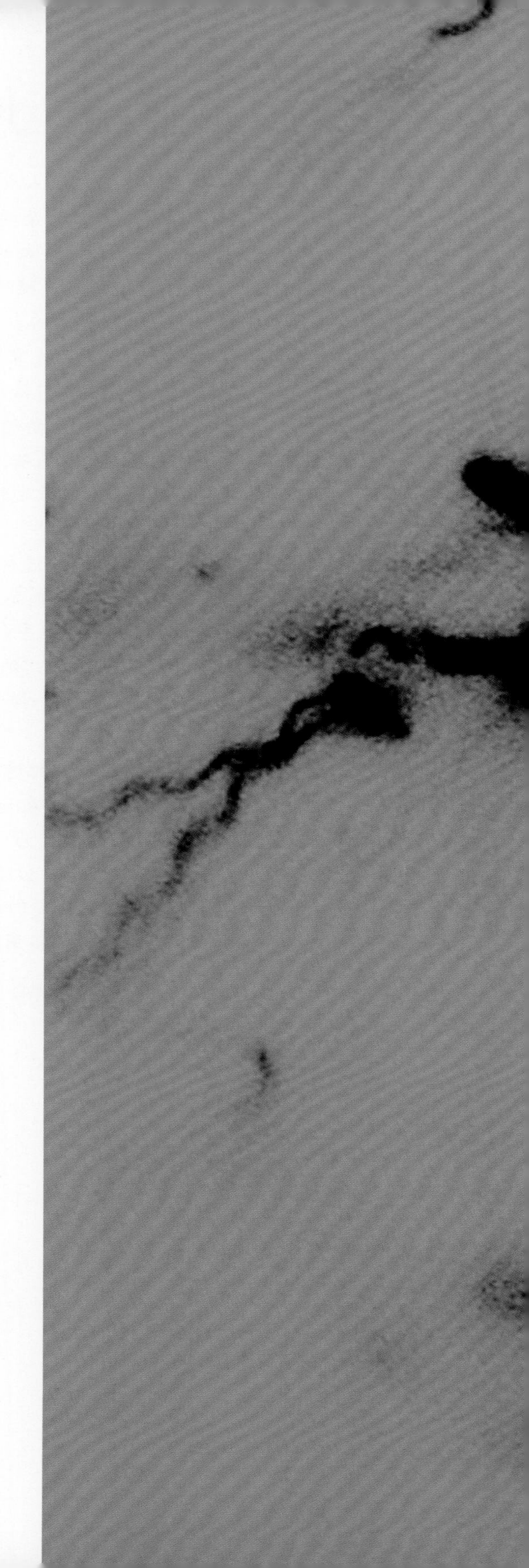

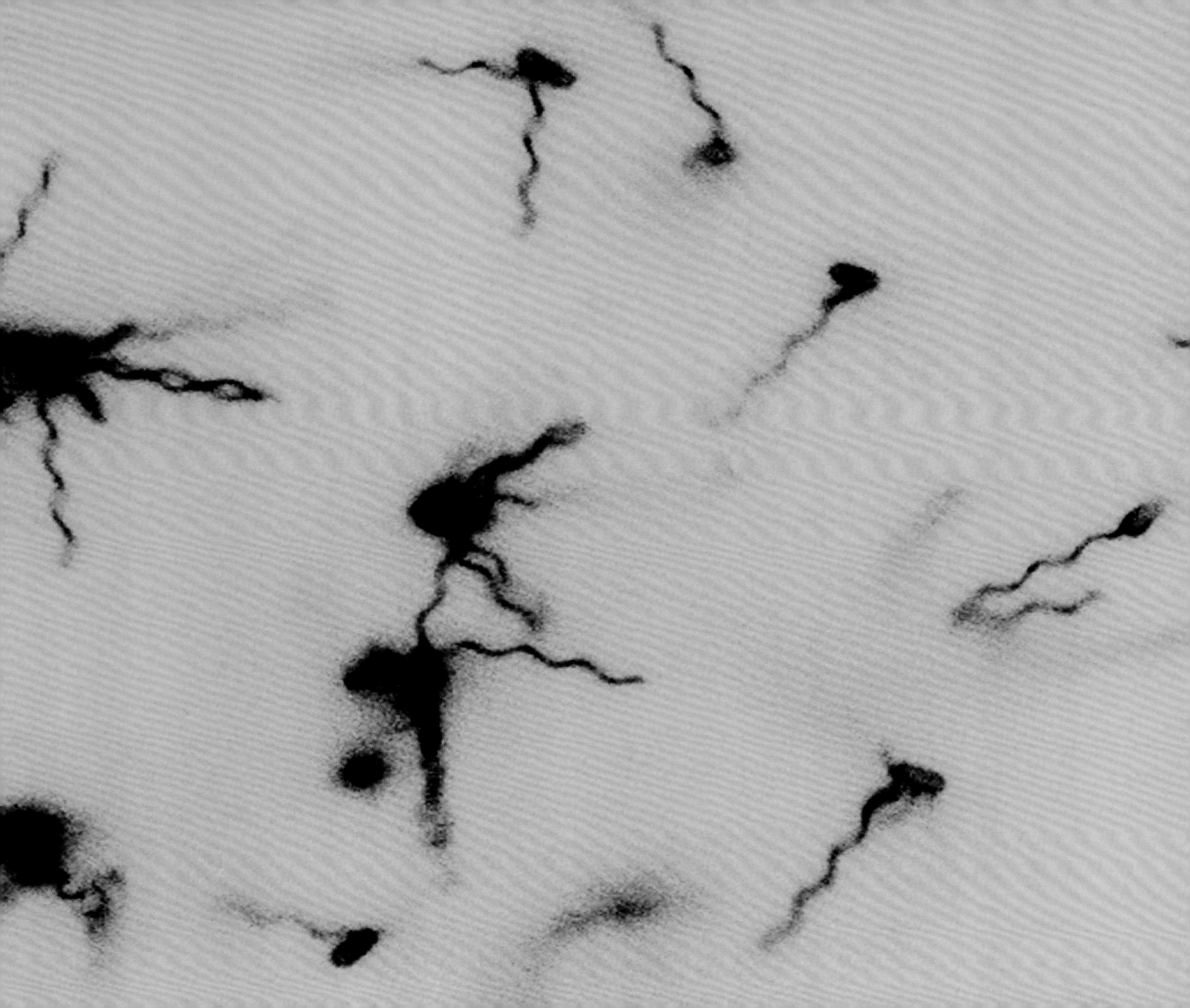

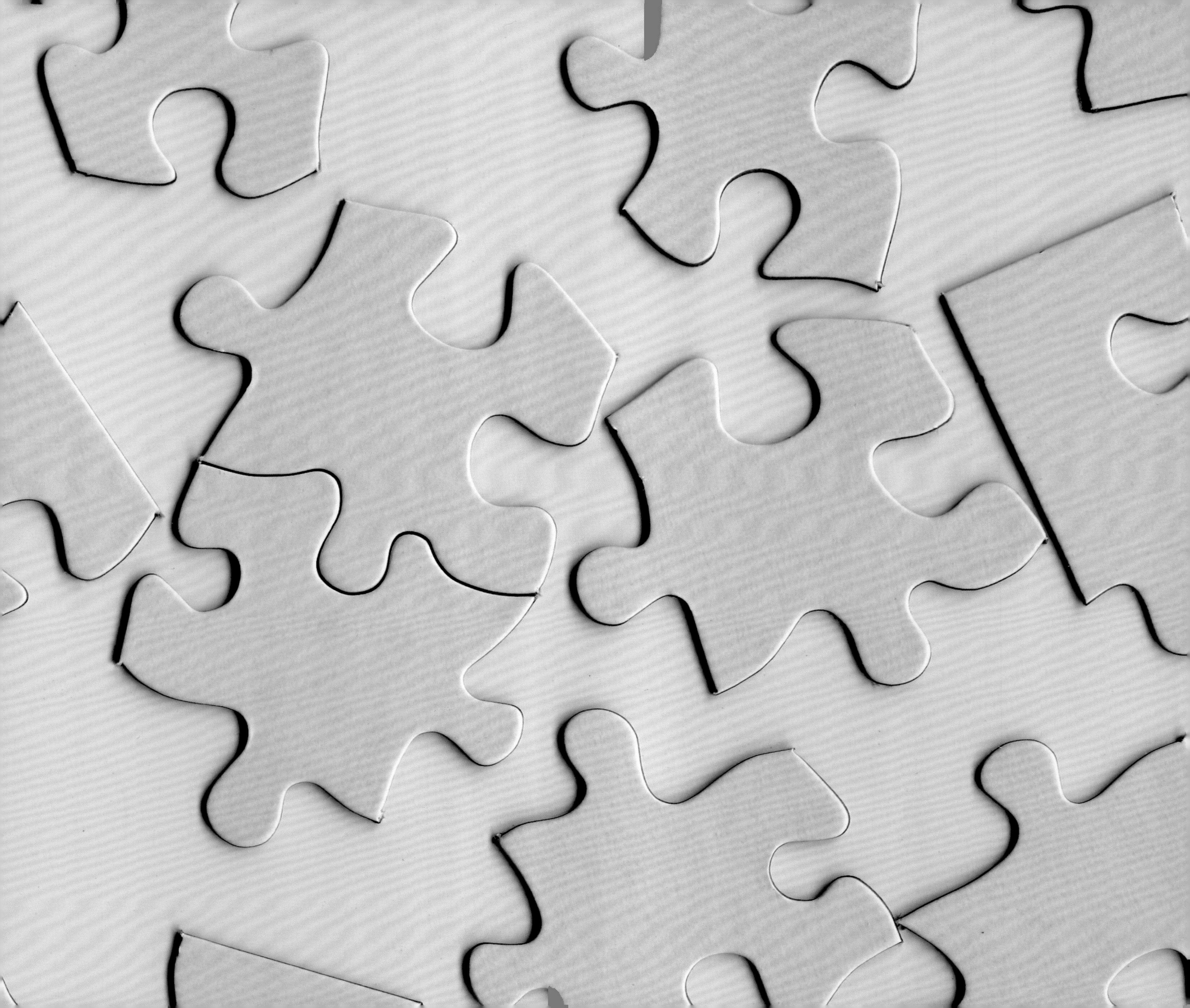

拼图般的生命

Life as a Jigsaw Puzzle

22

细胞的生命可以被简单形象地看作是一种拼图游戏。在这个游戏中,分子就是那些支离破碎的图块,当它们被正确地拼接在一起时,生命就诞生了。

然而,生命的诞生远比我们平时玩的拼图游戏更为复杂。孕育生命时所使用的大部分图块都不是平面的,而是三维、有弹性的,而且只有纳米大小。这就好比把果冻做成各种复杂的形状,而不是把硬纸板剪成各种简单的形状。要想把这些图块合理地拼接在一起,首先它们的形状必须匹配。就如同戴手套一样,我们可以把五只手指插进六指手套,而非四指手套。其次,在谜底为生命的拼图游戏中,我们不但要考虑图块形状的互补性,还要考虑其他因素。因为在这些图块的表面上,有些区域比较黏稠、有些比较光滑、还有的显示着不同的颜色,所有这些特征在拼接时都必须完全匹配。

上面的这些问题已经让这个特殊的拼图游戏相当复杂了,但实际情况还远不止如此。两块图块靠近时,哪怕已经匹配得相当完美,它们也不会呆在一起太久。事实上,这些图块是靠快速结合与分离来找到最佳匹配的。即便是在1秒钟内,它们都会与上千亿个"邻家女孩"碰面,寻求一见钟情并与之结合。令人不安的是,当解离时,这些图块可能已经改头换面了,它们所需的匹配对象也将随之而改变。因此,生命就如同一个多维、复杂的拼图,无数图块在不停地相互碰撞和转化,瞬息万变。

在这种拼图游戏中,细胞也不是吃素的,有时候它会吞噬并消化掉自己的兄弟姐妹,有时候则表演着分身魔术——把自己分裂成两个几乎完全相同的细胞。

在这个游戏中,我们还必须去注意另外一个重要的角色——那就是水。这些水分子充斥在各种图块之间,自由而快速地运动着,让人永远也猜不透它们的意图。有时它们是娇媚的红娘,在碰撞的过程中极力撮合图块的结合;而有时它们又变身为冷漠的巫后,无情地在图块间划出一道道永远无法逾越的鸿沟。我们知道水是生命之源,但在这个游戏里它究竟扮演着什么样角色呢?

生命仍然是一个难解之谜。

The simplest possible view of life—of the cell—is that it is a kind of jigsaw puzzle. The pieces are molecules. When they fit together correctly, life happens.

But there's more to it than that. First, imagine that most of the pieces are not flat but three-dimensional, rubbery, and only nanometers in size: Jell-O molded into complex shapes, rather than cardboard cut into primitive ones. To fit together, the shapes must be complementary: the hand must slide into the glove, and while a six-fingered glove might do for a five-fingered hand, a four-fingered glove would not. And "complementary" involves more than just shape—some patches of the surfaces are sticky, some slippery, and some colored, and these patches must also match.

All of which is complicated enough, but there is more. When the pieces come together, even when everything matches, they usually don't stick for long. (And incidentally, they are trying one another out for fit rapidly: every piece bumps its neighbors about a hundred billion times per second. Speed-dating in the fast lane.) Disconcertingly, when they separate, they may have changed into other pieces. So, if life is a jigsaw puzzle, it's a multidimensional puzzle of enormous complexity, with countless shape-changing pieces colliding and interconverting in a blur of activity.

And by the way, the puzzle also is carnivorous: feed it pieces of other puzzles, and it digests them into its own, and occasionally splits itself into two almost identical puzzles.

We've also left out the water: all of this molecular scurrying-around takes place in the company of other pieces—water molecules—which sometimes stick and get in the way, and sometimes act as glue or lubricant. We know that water is absolutely vital, but what, exactly, does it do?

We really don't understand how it all works.

As the Wheel Turns

转动的轮子

23

MOST OF THE MOTORS we build rotate a shaft around its axis. Rotary motors—electric motors and internal combustion engines—are perfect for turning wheels, and we are addicted to wheels.

What about bacteria? They do not have wheels, but they churn through water quite handsomely. What motor powers *them*? We infer from this ghostly cross section that bacteria also use rotary motors, but not like ours. The bacterial motor does have a rotating shaft, and a kind of stationary outer shell or bearing (shown here), braced somehow in the cell membrane. But its components are made of proteins—soft, intricately folded molecular threads—rather than hard, shaped metal. It runs on a current of protons rather than electrons; it turns a flagellum, not a propeller.

Yet it works! A bacterium at "full steam ahead" travels its own length more quickly than does a submarine. The bacterial motor is an inspiration for engineers: it is a fully formed, functional nanomachine, working in ways that we might never have considered. The bacterium pumps protons (H^+) out of its body and across its membrane, and thus creates a difference in their concentration between the inside and the outside. The protons can flow back into the cell by passing through the motor; when they do, the shaft ratchets forward, pushed by changes in the shape of its proteins. The process seems clunky—like pumping water uphill, so it can run down again to power a water wheel—but it is the one natural selection favored.

Nanotechnologists have argued fiercely about motors for decades. One of the early promises of the futurists was "little submarines": nanomachines circulating in the blood, hunting cancer cells. Micron-scale hunters *do* exist in our bodies (cells called lymphocytes), but they resemble a submarine no more than a cat resembles a lawnmower. Still, nanoscale flagellar motors propel micron-scale bacteria rapidly through water. These motors are wildly different from anything we have designed.

The flagellar motor is one of a flotilla of nanomachines with designs astonishingly alien to our intuition. But evolution has been throwing the dice—randomly testing designs—for millennia, with spectacular success. Among our tasks is to develop the humility and sophistication to learn from this master engineer.

人类制造的发动机往往离不开绕轴旋转的转子。发动机中绕轴旋转的转子、精密手表中的齿轮、电脑CPU上的小风扇……转轮的身影无处不在。

细菌没有转轮，却能在水中自由穿梭，这又是如何实现的呢？它们用的什么样的发动机呢？右图所示的正是细菌所用的神秘发动机的横截面。和我们日常生活中所用的发动机不同，它没有转子，取而代之的是一个固定在细胞膜表面类似于外壳或轴承一样的东西。它的各个元件都是由蛋白质构成的，这是一种柔软的、能任意折叠的线状物，而不是坚硬的、有固定形状的金属。更为迥异的是，这种发动机以质子流为驱动，而不是常规发动机所用的电子流。它靠驱动鞭毛提供前行动力，而不是螺旋桨。

这种发动机工作起来还是蛮棒的。如果把潜水艇缩小到一个细菌大小，你会惊讶地发现，全速前进的细菌会把潜水艇远远地甩在身后。在大自然的鬼斧神工面前，曾经让我们引以为豪的设计也会黯然失色。细菌中的天然纳米发动机就有着完美的设计工艺和超强的性能，其所采用的工作原理也是超乎我们的想象。细菌通过细胞膜把质子泵出体外，使细胞膜内外两侧的质子产生一定的浓度差；然后质子通过这个细菌发动机流回细胞内以达到内外浓度平衡。与此同时，发动机里的蛋白质也会产生形变，从而产生驱动力以推动整个细菌前行。整个过程如同用水车把水带到高处，再流下来以推动水轮，看似笨拙，其实是物竞天择的杰作。

纳米科学家已经在发动机这个话题上争论了数十年。科幻家则在很早以前就梦想着能驾驭微型潜水艇在血液里潜行以消灭癌细胞。事实上我们体内确实存在着微米级的、捕捉癌细胞的猎手，它就是淋巴细胞，但是把它比作潜水艇无异于把猫比作割草机。然而，纳米级的鞭毛发动机的确能驱动微米级的细菌在水里快速地游动，如此高效的发动机完全有别于人类所设计出来的机械系统。

对于大自然庞大的纳米机器"军团"来说，细菌神奇的鞭毛发动机只是一支毫不起眼的"部队"。自然界在进化时必须掷上数千年的骰子，随机地测试各种设计方案，最终才能得到这样的精品。而我们的任务就是谦虚认真地向大自然这个伟大的工程师学习。

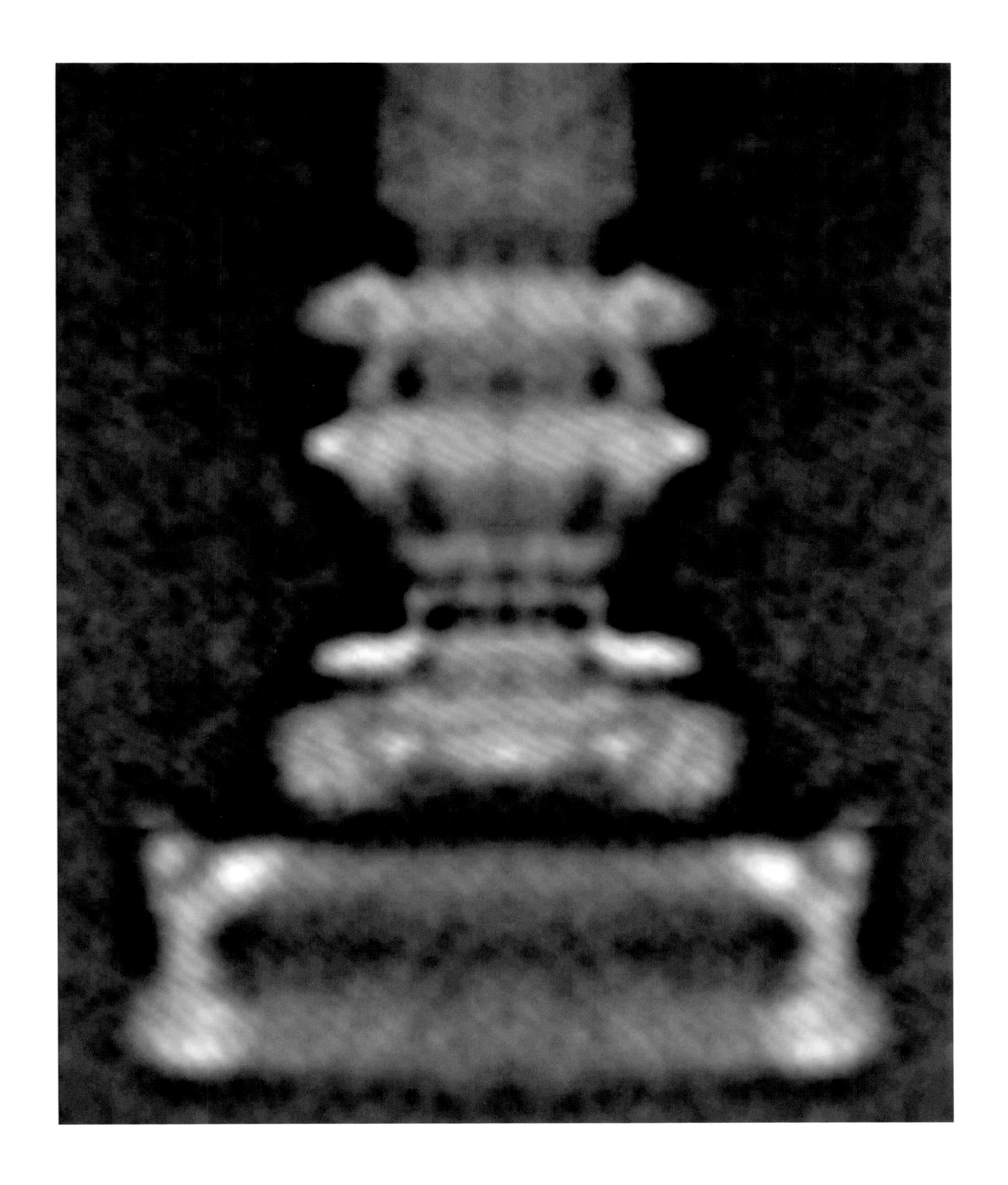

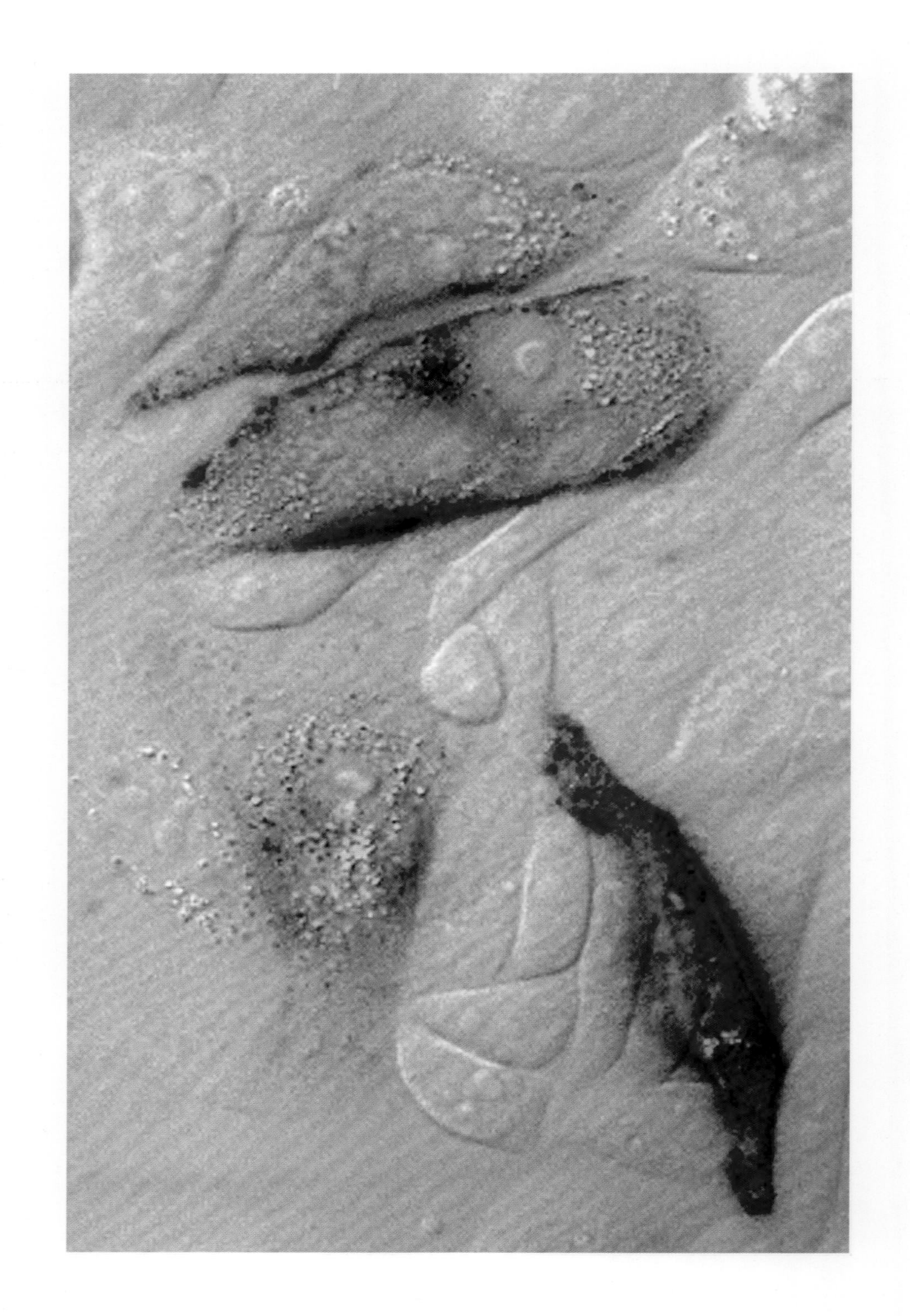

量子点和细胞

Quantum Dots and the Cell

24

在膝盖上轻轻敲一下音叉，它会发出弱弱的、单一频率的声音；重击它，音调不变，只不过声音更响而已。这个音调取决于音叉的振荡频率，与敲击它的力度无关。

虽然原理不同，但量子点（纳米级的无机球状材料，如硒化镉）的表现和音叉极其相似。当受紫外光照射时，它会发出特定颜色的可见光。增加照射光强度，其发射的可见光强度也会相应增加，但颜色不会改变。和音叉的机械振动不同，让量子点发出特定颜色光的振荡是来自其中的电子。

量子点的这个应用基于两个重要的性质。首先，它们发光的颜色由尺寸决定。尺寸较小的量子点偏向于发蓝光，较大的则偏向于发红光，如果尺寸处于两者之间，发射光的颜色则会介于红色与蓝色之间，例如黄光和绿光。我们可以借助化学合成的手段来精确地控制量子点的尺寸，进而有效地调控它们发光的颜色。其次，因为细胞内的大部分组织结构都是透明无色的，为了在显微镜下看清它们的内部结构，我们常常需要对其染色。许多染料分子在显微观察的过程中容易被损伤而失去发光的性能，量子点却不会出现这样的问题。

在泰国的夜间，萤火虫会聚集在某种树上并一起闪烁，于是这棵树的轮廓和内部枝状结构就被清晰而完整地勾画出来。该图显示从细胞内部量子点发出的荧光（叠加在细胞上的显微镜图像），不同的颜色来自不同尺寸的量子点发出的光，好似纳米级的萤火虫。通过对量子点表面进行修饰，可以选择性地勾勒出细胞内特定的结构。

WHACK A TUNING FORK GENTLY on your knee; it softly sings a single pure tone. A harder hit makes the same tone, just louder. This tone—that is, the frequency of the tuning fork's vibration—is independent of how hard it's hit.

Quantum dots (nanometer-sized spheres of inorganic materials such as cadmium selenide) behave similarly—although for a different reason. Shine ultraviolet light on a quantum dot, and it emits visible light of a particular color. Increase the intensity of the ultraviolet light, and the emitted light becomes more intense, but the color stays the same. The oscillations that produce the colors emitted by quantum dots are not the mechanical oscillations of a tuning fork, but the more subtle oscillations of their electrons.

Quantum dots are useful for two reasons. First, their *size* determines the color they emit: little dots emit in the blue; big ones in the red; medium-sized ones in between. It's easy to produce any color by just changing the size of the dot using clever synthesis. Second, to see the structures in mammalian cells by microscopy, we usually need to use dyes; most parts of a cell are transparent, uncolored, and invisible. Many dyes tend to bleach during examination in the microscope; quantum dots don't.

In Thailand, at night, fireflies collect in particular trees, and their flashing—sometimes astonishingly synchronized—shows the shape of the tree and the structure of its branches. This image shows colors emitted by quantum dots in cells (superimposed on a microscopic image of these cells). Different colors correspond to dots of different sizes. They have been tailored, like nanoscopic fireflies, to outline specific structures in the cell.

25 Sequencing DNA

DNA IS A CELL'S INSTRUCTION BOOK. Reading the letters and words in this book is no longer a problem, nor is it difficult to find the sentences. Understanding what a sentence *means*, and following the structure and plot of the book, remain largely beyond us.

The DNA tells the cell how to build proteins—the molecules that do its work. Proteins are made by stringing beads together; the beads are smaller molecules called amino acids, and they come in twenty different structures. The "alphabet" of DNA is simple: it has four letters, A, T, G, C—the "bases" of the genetic code. The "words"—the specifications of the next bead to add—are also easy: each is three letters. The "sentences"—the descriptions of complete proteins—are much more complicated: hundreds of words, which are sometimes mixed with sections that seem to be gibberish, sometimes scrambled with bits of other sentences, and sometimes extensively revised by unseen editors to change their meaning. With work, we can parse the sentences, but we really can't decipher what they say—what the proteins do. When we try to interpret the whole story—to understand "life"—we are like children facing *The Romance of the Three Kingdoms*. We suspect there is a plot, but we have no clue what it is.

Reading the story of the cell by reading its DNA is still a little like trying to understand a city by reading its telephone books. Telephone books give names, numbers, and addresses, but tell nothing about the people: what they do, who they talk to, what their favorite flavor of ice cream is, who works for government and who sells groceries. Nor do they tell how the people form a city. The DNA tells what some of the proteins are, but not what they do, how they communicate with one another, or (the *real* question) how the cell becomes "alive."

Reading DNA is fun and easy. Every biologist does it. The results are certainly instructive, and are becoming useful. Because DNA sequencing is so important, there are many technologies that do it rapidly and accurately.

DNA 测序

DNA 是细胞的指令书。如今,我们阅读这本书里的字母和单词已不成问题,甚至要找出其中的句子也算不上难事,但是想要弄明白每一个句子的含义并进一步理解这本书的结构和脉络,却仍是一个巨大的挑战。

DNA 指导细胞合成蛋白质。作为生命的基石,蛋白质就像一条珠链,而珠子就是二十种氨基酸分子。DNA 的"字母表"极为简单,它只用 A、T、C 和 G 这四个字母便可以构造所有的基因编码。每三个字母可以组成一个"单词",决定下一个要添加到链上的氨基酸。相比之下,"句子"可就复杂多了,它描述了整个蛋白质多肽链。不仅每个句子都包含几百个单词,有时还夹杂着一些胡言乱语或者不清不楚的话,也有时七零八落混乱不堪,有的甚至还会被那些看不见的编辑们(指基因变异)大幅度修改,从而失去了其原本的意思。经过不懈的努力,我们已经可以解析 DNA 的句子结构,但还无法破译每句话的含义,因此也就弄不明白蛋白质到底是怎么产生和工作的。当我们试着去解读生命奥秘的时候,就好似一个孩子面对《三国演义》这本书,也许能模模糊糊感觉到故事的情节,却找不出具体的线索。

通过解析 DNA 来解读细胞有点像通过翻阅电话号码簿去了解一座城市。号码簿上记载了众人的名字、电话号码和地址,但是对于他们是干什么的、他们平时跟谁打交道、他们最喜好什么口味的冰淇淋、谁为政府工作、谁卖杂货等问题,我们却一概不知。当然我们也不知道他们是怎么构成这座城市的。DNA 给我们提供了大部分蛋白质的信息,却没有告诉我们这些蛋白质的用处和相互关系,更不要说"细胞是如何运作的"这种深奥的问题了。

解析 DNA 编码既简单又充满乐趣,几乎每个生物学家都在做这种事。所得到的结果正在逐步发挥作用。现如今,有许多快速而又精准的技术可以来协助我们完成这项重要工作。

WHITE PAGES

CAATGGTTCAATGGTTGGTCAAGGATGCGCAACCCAATGATTCTTTGTTCCTTCATTATTCT
GGTGGCCAGGTGGCCAAACTGAAGATTTGGATGGGGACGAAGAAGATGGGATGGATGATGTTATATAT
CCGGTCGACCGGTCGATTTCGAAACTCAAGGGCCAATTATCGACGATGAAATGCACGATATAATGGTG
AAGCCCTTAAGCCCTTACAACAAGGTGTTAGACTAACAGCATTGTTTGACTCTTGTCATTCGGGTACA
GTGTTGGAGTGTTGGATCTTCCATATACCTATTCTACTAAGGGTATTATTAAGGAGCCCAATATTTGG
ATTCTCAGATTCTCAGC.GAACGCGTGAGACAGATCAAATTCTCAGCAGCAGATGTTGTTATGTTATCA
TGCTGTCGTGCTGTCGA.GATAATCAAACTTCTGCAGATGCTGTCGAAGATGGGCAAAATACAGGTGCA
TTTACAACTTTACAACC.GCCTTCATCAAGGTTATGACTTTACAACCACAGCAATCATATTTATCTCTT
ATGTATCCATGTATCCAGGTAGTGGACGTTACACCTACAACAACGCTGGTGGTAATAATGGCTACCAA
CGGCCCATCGGCCCATGGCTCCTCCACCTAACCAGCAGTATGGACAGCAATATGGTCAGCAATATGAA
CAGCAGTACAGCAGTATGGACAGCAATATGGGCAACAAAATGATCAGCAATTCAGTCAGCAATATGCT
CCACCACCCCACCACCAGGTCCTCCCCCTATGGCTTATAACAGGCCTGTGTATCCCCCCCTCAATTC
CAGCAGGACAGCAGGAACAGGCAAAGGCACAATTAAGCAACGGCTACAACAATCCTAATGTAAACGCA
TCCAATATTCCAATATGTACGGTCCACCCCAGAATATGTCATTACCTCCACCTCAAACACAAACTATT
CAAGGTACCAAGGTACAGACCAACCTTATCAGTATTCTCAATGTACTGGGCGTAGAAAGGCTTTGATT
ATCGGTATATCGGTATAAACTACATAGGTTCAAAAAATCAACTGCGTGGTTGTATCAATGATGCTCAT
AACATCTTAACATCTTCAACTTTTTGACTAATGGGTACGGTTACAGTTCAGATGACATTGTCATATTA
ACTGATGAACTGATGATCAGAACGATTTGGTCAGGGTTCCCACTAGGGCTAATATGATTAGGGCCATG
CAATGGTTCAATGGTTGGTCAAGGATGCGCAACCCAATGATTCTTTGTTCCTTCATTATTCTGGACAT
GGTGGCCAGGTGGCCAAACTGAAGATTTGGATGGGGACGAAGAAGATGGGATGGATGATGTTATATAT
CCGGTCGACCGGTCGATTTCGAAACTCAAGGGCCAATTATCGACGATGAAATGCACGATATAATGGTG
AAGCCCTTAAGCCCTTACAACAAGGTGTTAGACTAACAGCATTGTTTGACTCTTGTCATTCGGGTACA
GTGTTGGAGTGTTGGATCTTCCATATACCTATTCTACTAAGGGTATTATTAAGGAGCCCAATATTTGG
AAGGATGTAAGGATGTTGGCCAAGATGGCCTGCAAGCAGCTATTTCATATGCCACAGGAAACAGGGCT
GCTTTGATGCTTTGATTGGTTCTTTAGGTTCTATATTCAAGACCGTTAAGGGAGGTATGGGCAATAAT
GTGGATAGGTGGATAGAGAACGCGTGAGACAGATCAAATTCTCAGCAGCAGATGTTGTTATGTTATCA
GGTTCGAAGGTTCGAAGGATAATCAAACTTCTGCAGATGCTGTCGAAGATGGGCAAAATACAGGTGCA
ATGTCCCAATGTCCCACGCCTTCATCAAGGTTATGACTTTACAACCACAGCAATCATATTTATCTCTT
TTACAGAATTACAGAACATGAGGAAAGAATTGGCTGGTAAGTATTCTCAAAAACCACAATTATCATCG
TCACACCCTCACACCCTATTGACGTAAATCTGCAATTTATTATGTAGTCACACCCTATTGACGTAAAT
CAACAAGGCAACAAGGTGTTAGACTAACAGCATTGTTTGACTCTTGTCATTCGGGTACAAGCATTGTT
CTTCCATACTTCCATATACCTATTCTACTAAGGGTATTATTAAGGAGCCCAATATTTGGTAAGGGTAT
GGCCAAGAGGCCAAGATGGCCTGCAAGCAGCTATTTCATATGCCACAGGAAACAGGGCTAGCTATTT
GGTTCTTTGGTTCTTTAGGTTCTATATTCAAGACCGTTAAGGGAGGTATGGGCAATAATCAAGACCG
GAACGCGTGAACGCGTGAGACAGATCAAATTCTCAGCAGCAGATGTTGTTATGTTATCAATTCTCAG
GATAATCAGATAATCAAACTTCTGCAGATGCTGTCGAAGATGGGCAAAATACAGGTGCATGCTGTCG
GCCTTCATGCCTTCATCAAGGTTATGACTTTACAACCACAGCAATCATATTTATCTCTTTTTACAACT
ATGAGGAAATGAGGAAAGAATTGGCTGGTAAGTATTCTCAAAAACCACAATTATCATCGTAAGTATT
ATTGACGTATTGACGTAAATCTGCAATTTATTATGTAGTCACACCCTATTGACGTAAATTATTATGTA
AGCATTGTAGCATTGTTCAACAAGGTGTTAGACTAACAGCATTGTTTGACTCTTGTCATTCGGGTACA
TAAGGGTATAAGGGTATCTTCCATATACCTATTCTACTAAGGGTATTATTAAGGAGCCCAATATTTGG
AGCTATTTAGCTATTTCGGCCAAGATGGCCTGCAAGCAGCTATTTCATATGCCACAGGAAACAGGGCT
CAAGACCGCAAGACCGTGGTTCTTTAGGTTCTATATTCAAGACCGTTAAGGGAGGTATGGGCAATAAT
ATTCTCAGATTCTCAGCGAACGCGTGAGACAGATCAAATTCTCAGCAGCAGATGTTGTTATGTTATCA
TGCTGTCGTGCTGTCGAGATAATCAAACTTCTGCAGATGCTGTCGAAGATGGGCAAAATACAGGTGCA
TTTACAACTTTACAACCGCCTTCATCAAGGTTATGACTTTACAACCACAGCAATCATATTTATCTCTT

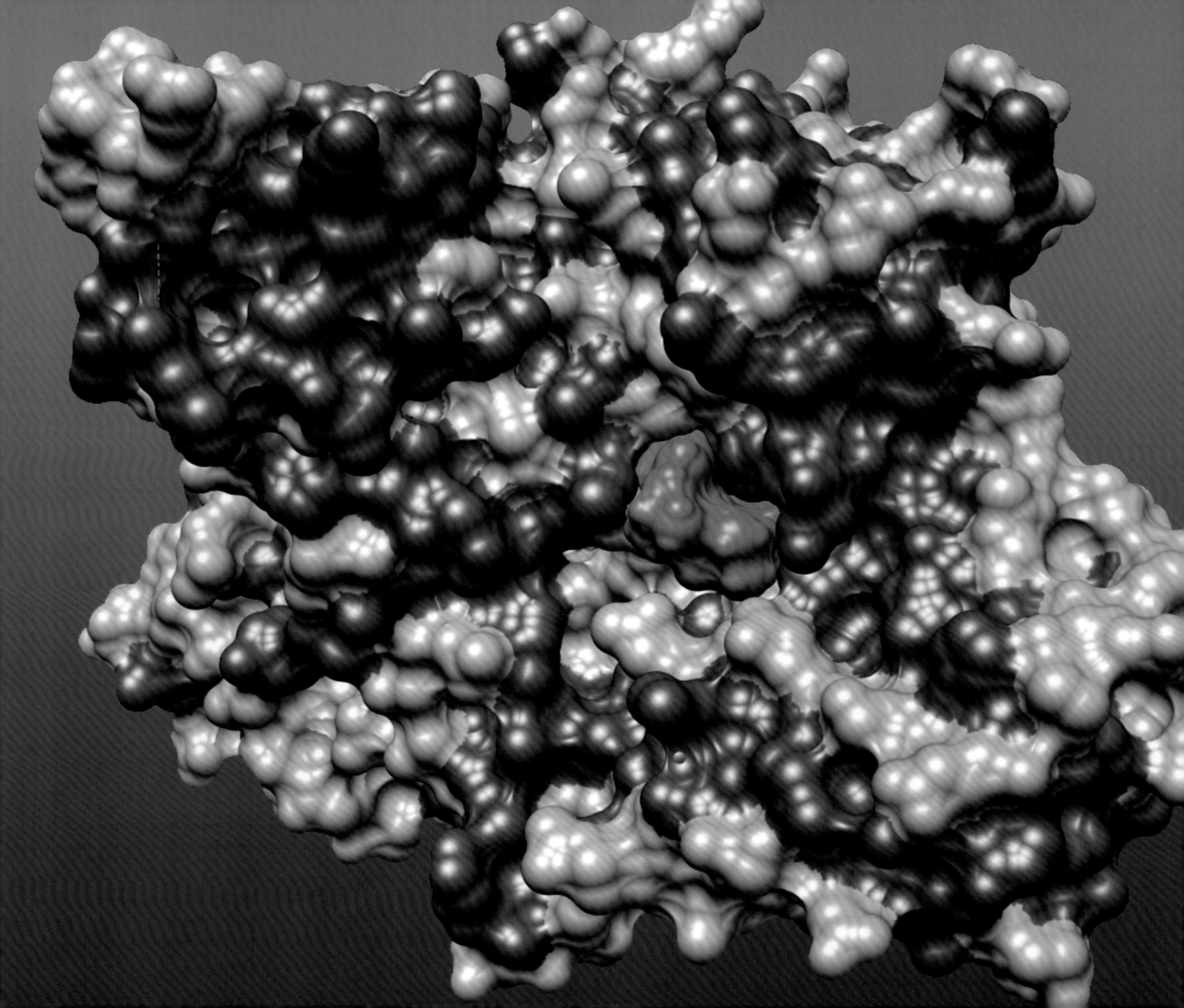

分子识别

Molecular Recognition

26

小孩对怎样在一场聚会中找到走散的父母有着很敏锐的知觉。他们能对成千上万双膝盖进行分类，找到形状熟悉的那一对，然后将自己的身体挤进它们中间。同时，小孩也擅长改变大人们的不良习惯。大人通常只关注大人，但有小孩在场时，他们则会对小孩更加上心。

分子在细胞中做着同样的事。它们找到合适的分子时会相互结合，一个在周围环绕着，另一个则被楔在里面。如果它们做的对的话，细胞便能存活；否则，细胞则会凋亡。这一切是如何实现的呢？分子是如何识别对方的呢？是什么把它们拴在一起的呢？

分子是有形状的，形状的互补与配对是分子识别的基础与准则。当然，除了形状，这其中还有很多其他要素在起作用。就好比一只让人熟悉的手，不只是五根手指，还包括其他细节，比如掌纹、力度、温度、老茧。

对于一个分子来说，其表面可以是带电荷的也可以是中性的，可以是亲水的抑或是疏水的，柔性的或者是刚性的……最重要的是，它总是被水包围着。分子和水之间的相互作用是分子识别的关键所在，但时至今日，我们对它们之间的具体作用机制仍然是一知半解，往往是知其然而不知其所以然。

这张图片所展示的是用电脑合成出的一种直径约为几纳米的蛋白质分子的表面，它对体内胆固醇的合成至关重要。血液中过多的胆固醇会造成动脉梗塞，后果不堪设想。图中蓝色的分子是一种叫做抑制素的药物，这种药物能减缓体内合成胆固醇的速度。若没有每天服用这种抑制剂去正确地识别蛋白质并且插入到蛋白质的活性中心，许多老人都会过早地离世。

分子识别是生物化学中最基本的过程。从正确识别DNA中的信息，到制造完美折叠的蛋白质，直至调控生命的新陈代谢，细胞中每一步反应无不依赖于这一过程。我们甚至可以说，分子识别之于细胞的功能比父母之于孩子的成长更为重要。

LITTLE CHILDREN HAVE AN unerring sense for how to find their parents at a party. They sort through the forest of knees until they find the pair with the right shape, and then wedge themselves between them. Children are very successful in doing what they set out to do: to block a dysfunctional behavior (grownups paying attention to grownups) and cause more useful behavior (parents paying attention to children).

Molecules in the cell do the same. They find one another and stick together—one wraps *around*, and the other wedges *in*. If they do it correctly, the cell lives; if they make mistakes, the cell dies. How does it work? How do they find one another? What holds them together?

Molecules have shapes, and the fitting together of complementary shapes—*molecular recognition*—is part of the answer. But there is more to it than simply shape. A familiar hand is not just five fingers; it is also texture, and strength, and warmth, and calluses.

So it is with molecules. Some parts of their surfaces are electrically charged; some attract water, some repel it; some are rubbery; some are stiff. And importantly—perhaps *most* importantly—they are surrounded by water. We know that water, and the interactions of water with the molecules dissolved in it, are key to molecular recognition, but we do not understand how.

The large textured surface in this image is a computer's view of a protein (a few nanometers in diameter) indispensable to the synthesis of cholesterol. Too much cholesterol, and our arteries plug, with dire consequences. The small blue object is a drug called a statin, a drug that slows the rate of synthesis of cholesterol in the body. Many older people would be dead had not the statins they take daily found the right proteins and wedged themselves between the proteins' knees.

Molecular recognition is perhaps the most fundamental single process in biochemistry. Everything in the cell depends on it—from reading the information in DNA correctly, through making properly folded proteins, to regulating metabolism. It is even more fundamental to the functioning of the cell than parents are to the functioning of their children.

27 Harvesting Light

捕 获 光

As BEETHOVEN AGED, he grew deaf—a serious problem for anyone, but especially for a musician. An ear trumpet was a partial solution. It was a tuba in reverse: the wide end collected sound, concentrated it, and funneled it to his fading ear. Scratchy whispers become muffled shouts.

Plants have a problem related to Beethoven's. Sunlight is vital to them—the food on which they live—and sunlight is dilute. What is bright to our eyes may not be bright enough for a plant. It uses a solution similar to Beethoven's. Its leaves contain marvelous, complex structures of molecules that are—in essence—ear trumpets for light. These circular structures absorb red light (and reflect green, so we perceive leaves as green) and funnel the energy from the absorbed light to even more complex structures at their center. These then begin the intricate, multistep process of generating oxygen from water, and synthesizing from carbon dioxide the sugars the plant uses to build itself.

Those of us who love Beethoven's music benefited from his ear trumpet (the one in this image is *his* trumpet—the one Beethoven himself used). Everyone—including Beethoven—has depended on the "light harvesting trumpet" of plants, since it makes possible all that we eat and all that we breathe.

晚年的贝多芬双耳失聪,这样的状况对于任何人来说都是一个沉重的打击,特别是对一位音乐家而言。类似于号角的助听器能帮助他解决一部分听力问题。这种助听器像一把反向的大号:广角端收集声音,然后把声音集中起来,经过貌似漏斗的听筒传送到听力衰退的内耳。这样,沙哑的耳语就能变成低沉的呼喊声。

植物也有和贝多芬一样的困扰。阳光对植物来说举足轻重,因为植物依赖阳光而生存。然而,太阳光非常分散,即便是对于我们的肉眼来说已经很刺眼,对植物来说也算不上充足。在这样的情况下,植物是怎么做的呢?它们利用了和贝多芬使用助听器类似的方法来捕获分散的阳光。植物的叶子中含有一些神奇的、结构复杂的分子,它们能像助听器放大声音一样有效地捕获太阳光。这些环状的分子能吸收红光,反射绿光,所以我们看到的叶子大多是绿色的。然后,它们将捕获的光能输送到结构更为复杂、处于叶子中心的分子。自此,错综复杂的光合作用便拉开了序幕:消耗水和二氧化碳,产生氧气以及供植物生长所需的糖分。

对于喜爱贝多芬的音乐的人来说,他的助听器(图中所展示的正是贝多芬当年用过的助听器)使我们受益匪浅。而对所有人而言,包括贝多芬在内,都应该感谢植物的"光捕获器",正是它为我们创造了赖以生存的食物和氧气。

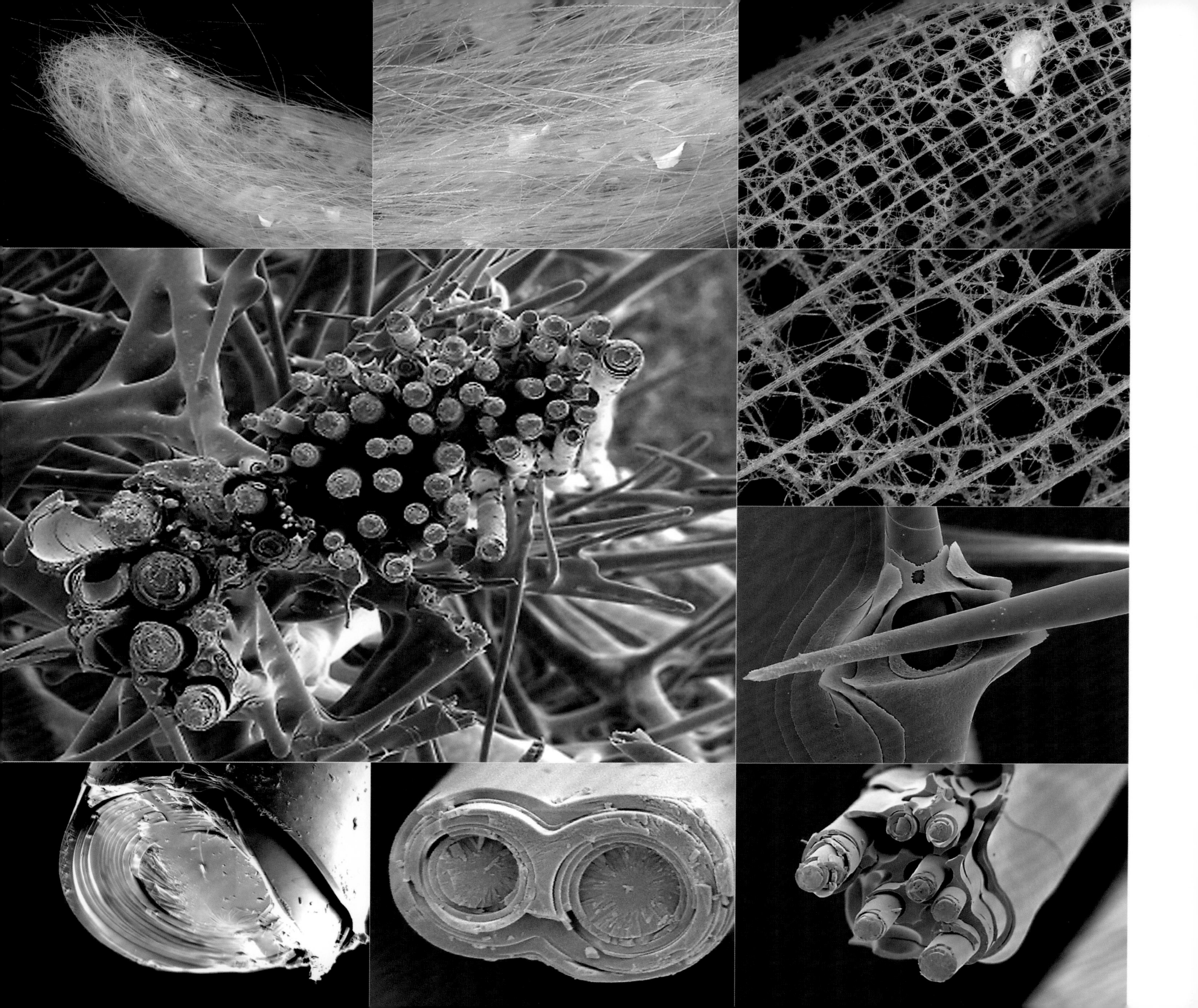

低等动物的优雅

The Elegance of Simple Animals

28

海洋里生活着许许多多很复杂的生物，我们通常只注意到那些智慧型的，如乌贼、海豚和鲸鱼。在某些方面，它们可以轻易地超越我们人类，比方说在三维空间下棋，在整个海洋范围内进行大合唱，还有用绝妙的办法来躲避鲨鱼。它们那敏锐的警觉性总是让我们望尘莫及，即便它们偶尔也会成为我们盘中的美餐。

再来看看那些貌似不怎么聪明的生物是怎样的吧。它们或许没有大脑，没有眼睛，不会移动，更不像迪斯尼动画里描绘的那样活泼可爱，但它们就没有值得我们欣赏的技能吗?

下一页的图片展示的是一个叫做维纳斯花篮(曲莓菌)的海绵骨架。这个生物给自己建造了一个精致的二氧化硅多层牢笼(活着时依附在这个骨架上的细胞已经不见了)。一对雌雄小虾为了逃避危险，躲进了牢笼，用自由换来了安全，直到终老。海绵骨架和小虾们的关系极其微妙，笼子的确保护了小虾，但小虾也给骨架上的细胞提供了食物，它们谁也离不开谁。

仔细观察，我们会发现笼子的每个部分都由更小的单元组成，而在不同尺度的单元中，都有相互交织的周期性结构，其构造细节真可谓是巧夺天工。例如，那些在海绵组织底部的支柱——骨针，既能提供足够的机械强度用以支撑整个结构，又能传导附生的细菌所发出的磷光，这犹如闪烁在加州黑店门口的霓虹灯，它似乎在招引顾客:“欢迎随时入住，来了就别想走啦”。

人类的任何建筑设计都无法与这个海绵骨架相媲美。除了海水以外，海绵骨架并没有任何其他的东西可以利用。它甚至连所谓的大脑也没有，但却设计并制造出了如此巧夺天工的艺术品，这就足以证明大自然的精湛技艺。在它的鬼斧神工面前，人类只能甘拜下风。

好在我们不食用这个精致的艺术品!

IN THE OCEAN LIVE CREATURES of many complexities. We focus our admiration on the intellectuals: squid, porpoises, whales. They are the ones who would—in a different world—easily beat us at three-dimensional chess or oceanwide choral singing, and already do a better job of evading sharks. We are amazed at the subtlety of their sentience. We sometimes enjoy eating them.

But what about the creatures that are less smart? The ones with no brain, no eyes, no mobility, no Disney-ready personality? Have they no tricks worth our admiration?

The image on the following page is the skeleton of a sponge called Venus's flower basket (*Euplectella aspergillium*). This blob of a creature builds for itself an intricate, hierarchical cage of silica (here, the cells that covered this structure when the sponge was alive are gone). In this cage, the sponge traps two shrimp (he and she), who spend their lives there, both protected and imprisoned. The relationship between the sponge and the shrimp is complicated, as are most *ménages à trois*. The cage seems to protect the shrimp; the sponge and shrimp somehow help one another to feed.

Each element of the cage is composed of smaller elements, and at each scale of size there are astonishing and unexpected details of design, construction, and interweaving of elements. For example, the struts—the "spicules"—simultaneously provide mechanical strength and transmit light generated by phosphorescent bacteria in the base of the organism to its rim—perhaps the neon lighting of an abyssal Hotel California for passing prey. "You can check out any time you like, but you can never leave."

This design is as elegant as anything engineers can conceive, and its fabrication—particularly since the sponge has nothing even faintly resembling a brain, and no raw materials to work with other than seawater—is beyond the best that engineers can do.

We do not eat it.

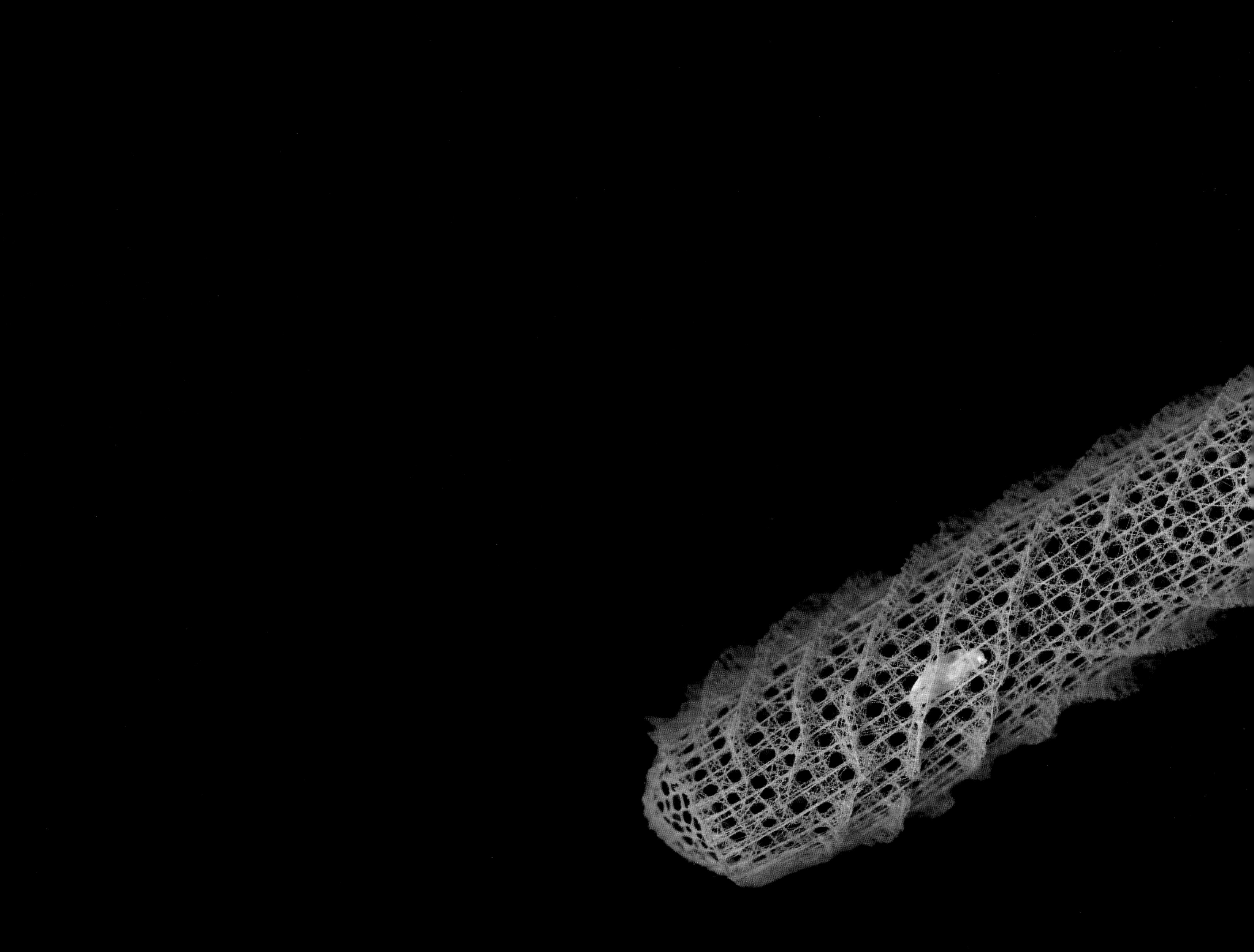

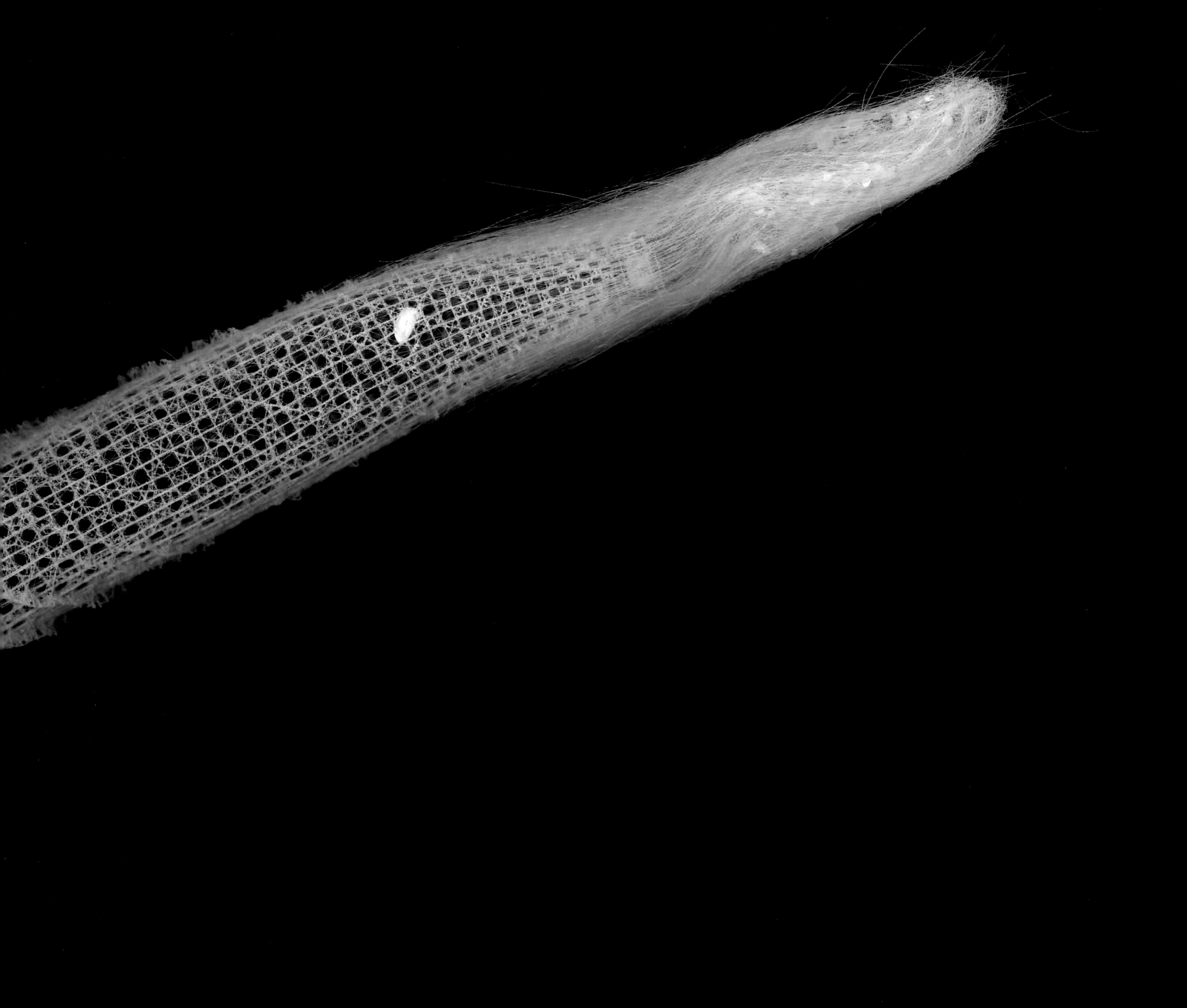

Antibodies

抗 体

29

The cute little two-year-old with the fever—the one doing her bit to enable her virus to achieve its manifest destiny—has succeeded. You have the flu. Now that you are infected, what do you do about it? Before the virus destroys your lungs, and—quite incidentally for the virus, if not for you—kills you?

We have evolved a clever strategy to protect ourselves against infectious two-year-olds. We deploy molecules called antibodies that attach themselves to anything—proteins, cells, viruses—that is "not us." If it is not recognizable as "friend," it is assumed to be "enemy." Antibodies are a little like the family pit bull: loving to the children; hostile to all strangers. Once the antibodies are attached, the stranger is labeled as dangerous, and the adults—cells called macrophages and killer T cells—attend to him. To make sure their grip on the stranger is firm, antibodies come with two "hands."

Are there mistakes? Yes. "Allergies" are this wonderful system getting it wrong; the cells that produce antibodies can mistake bee venom, pollen, proteins in fish soup, or feces from dust mites as evidence of infection by parasites, with unhappy consequences. Autoimmune diseases such as arthritis, lupus, and myasthenia gravis also show the immune system mistaking "friends"—our own tissues—for "enemies"—pathogens.

Scientists often learn by combining bits of information from many sources, as pointillists painted using many dots of color. This image—antibodies spread on a surface and examined by electron microscopy—looks, at first glance, more like the noise on a TV screen after hours than anything else. But look closely: with a little imagination, some of these blobs can be interpreted as Ys—or rings of Ys—a few nanometers in size. These shapes—even seen only fuzzily—confirm an aspect of the structure of antibodies: that they have two "hands" and one "body." It's just a dot of color on the canvas, but it helps to paint the picture.

两岁的小家伙正发着烧，因为得了流感。十有八九，你很快也会感染上同样的病毒。在病毒摧毁你的肺或杀死你之前（当然，只要你能挺得住，病毒导致死亡的可能性还是很小的），你的身体会做些什么呢？

人类已经进化出了一套强有力的防御系统，用来保护自己尽量不被病毒感染。通常，我们靠指挥一种叫做抗体的分子来对付入侵者，它可以和任何敌对势力相结合，例如外来蛋白质、细胞和病毒等，从而使敌军丧失战斗力。如果入侵者不能被抗体识别为友军，那么就会被其默认为敌军。抗体有点像家养的斗牛犬，它们爱护自家孩童，敌视所有陌生人。一旦抗体与陌生人相结合，这个陌生人就会被贴上危险的标签。同时，巨噬细胞和杀伤性T细胞就会注意到它们并防范和对付这些危险人物。为了牢牢地抓住陌生人，每个抗体都有两只手臂。

那么，抗体会犯错误吗？答案是肯定的。过敏反应便是这个神奇的防御系统犯迷糊时所造成的。产生抗体的细胞可能会把诸如蜂毒、花粉、鱼汤中的蛋白质，甚至尘螨的排泄物当作被寄生虫感染的信号，从而诱发不该有的战争，让人体产生不良反应。自身免疫病如类风湿性关节炎、系统性红斑狼疮和重症肌无力等，都是免疫系统错误地把自身组织当成了病原体一样的敌人所致。

科学家们经常会从各个方面获取信息并加以整合，进而从不同角度去学习、研究和揭示真理，就如同点彩派画家使用丰富的色点绘画一样。右图是抗体的电子显微照片。乍一看，分散开的抗体好似电视机屏幕上的雪花噪点。但是如果仔细观察并稍微发挥一下想像力，你就会发现，这些团状物似乎是许多Y字型结构或Y字组成的环状构成，其大小仅有几个纳米。尽管不算特别清晰，但这张图片也足以说明抗体除了有“身体”以外，确实还有一双小“手臂”。这个发现似乎微不足道，但对我们理解抗体是如何运作的起了关键的作用。

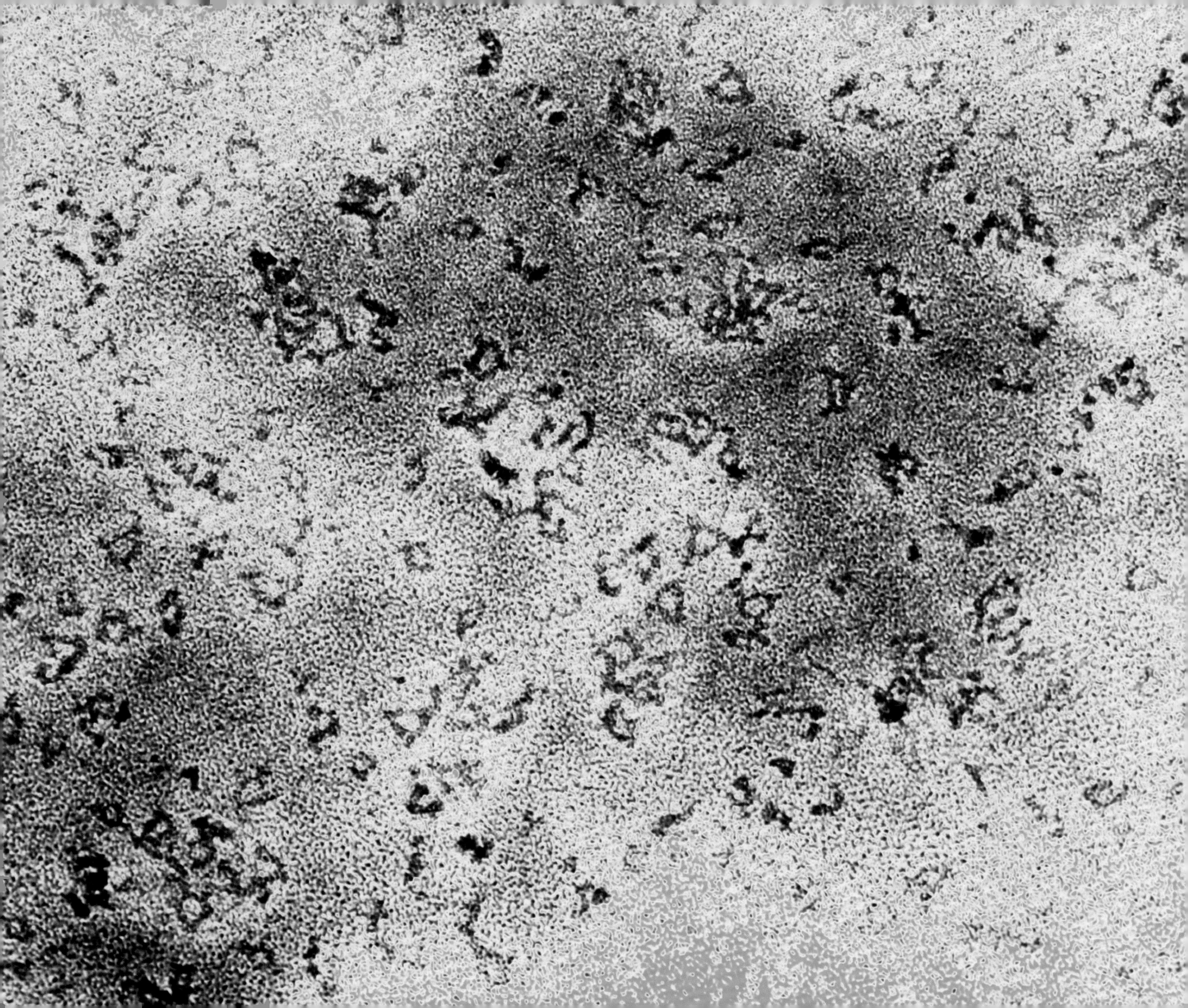

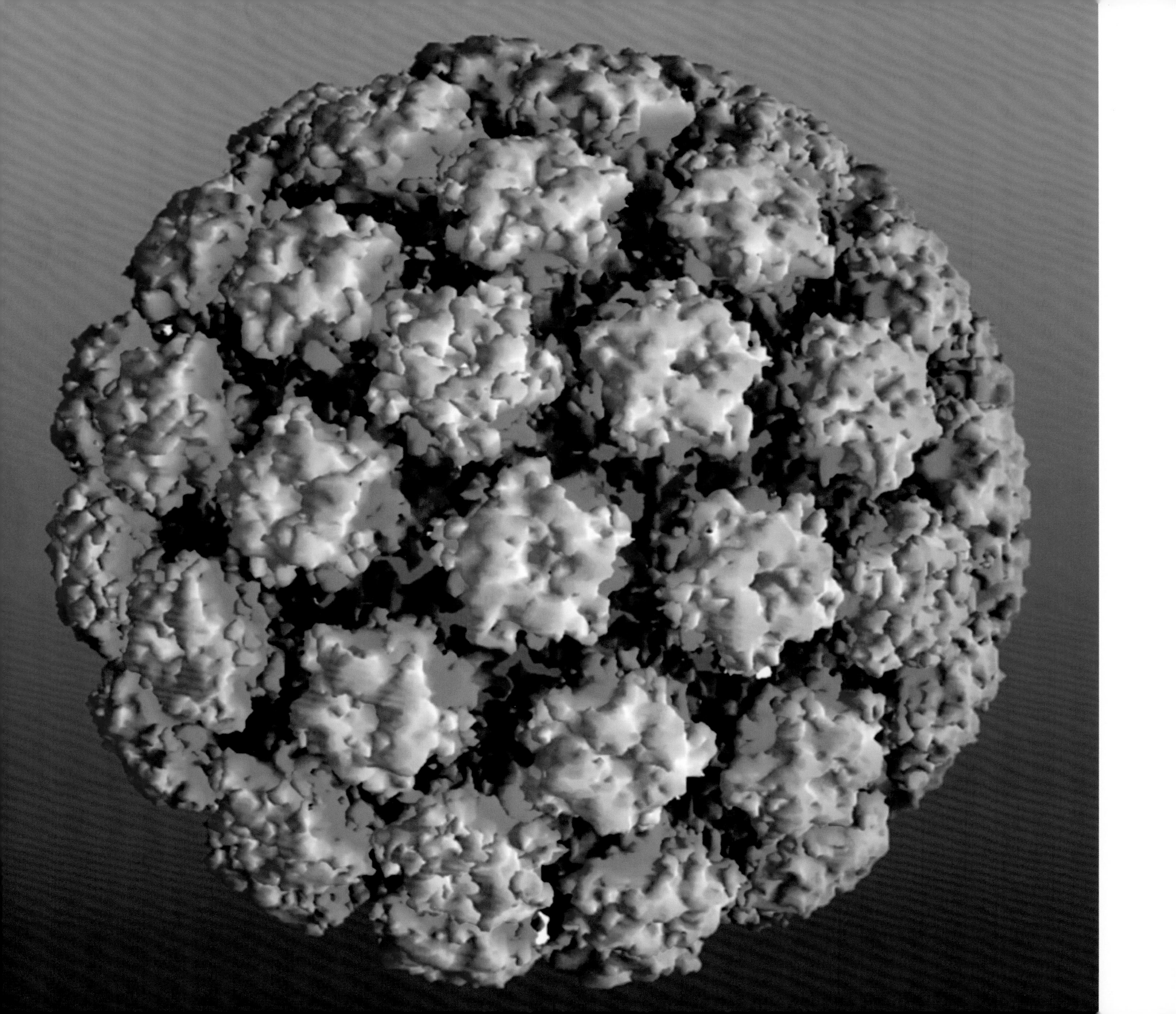

病　　毒　　Virus

30

我们用文字来思考。没有准确的文字描述，我们的思维将会变得苍白无力。看到“活着”这个词，我们会不经意地想起那些活着的东西，并会构思怎样去描述它们。然而，更多的东西并没有活着。在描述它们的时候，我们也许会惊讶地发现自己的词汇是多么的匮乏。例如，我们该怎么去描述石头呢？“不是活着的”“不能动弹的”“没有生命的”……这些描述都是说石头不是这样那样的，而没有说明白石头到底是什么或者是怎样的。“死亡”的潜台词是“曾经存活过”。我们似乎有这样的盲点：某种东西要么活着或曾经存活过，要么它就什么都不是。

如果只有“活着”和“死亡”这两个基本的词汇，我们确实可以在两者之间划出一条明显的界线。但是事情真的如此简单吗？难道“活着”和“死亡”就像“白”与“黑”那样泾渭分明？两者之间就没有一个过渡地带？

我们知道地球在40亿年前诞生了生命，但并不清楚那些无生命的东西最初是如何变成有生命的物体的。我们不知道在生命诞生之前，那些构造生命的分子是如何学会生存的。虽然从无生命到有生命并不是一个可以严格定义的过程，但是那些孕育生命的物质在成为生命之前到底是什么我们真的无从知晓。我们只知道，现今的生命总是源于生命——细胞分裂、鱼产卵繁殖，或者婴儿的诞生。

相比之下，我们对生命由生到死的过程似乎了解得更多更完善，因为我们自己总有一天要死亡。我们知道死亡是一点点逼近的，但却无法确认一个动物或细胞是在什么时候死亡的。我们该如何描述死亡呢？也许我们需要一个词来连接“生”和“死”这两个状态。活着？部分活着？身体机能还在运作的死亡？还是彻底的死亡？

病毒游走在生与死的边界上。在细胞外，它是一个装着DNA（或RNA）的包裹，也可以说是一小簇没有生命的分子。但当它遇到一个合适的细胞时，就会与之粘附，并把自己的DNA注入其内。被注入的病毒DNA会奴役整个细胞，打乱细胞的正常机能，并迫使细胞制造出更多的病毒。所以，在细胞外，病毒是死的，而一旦进入到细胞内，它就会变成生命的一部分。

那么，我们到底该怎样描述病毒呢？无生命的物种、有潜在生命的活种、偶尔部分有生命的东西、能控制生命的非生命物质，还是其他什么东西？如果我们能接受某些东西如病毒、干燥的孢子、冷冻的胚胎等既非活着也非死亡的观念，认同有一种状态介于生死之间的观点，那么我们的世界将变得更加复杂。

对活着的东西要尽道德义务，这是我们的底线。但是对于那些半死不活的，或者说是似死似活的东西，我们应该尽到什么样的义务呢？从一个崭新的角度去思考生命和近乎生命的东西，我们需要新的观念和词汇。

WE THINK IN WORDS. Without accurate words, we think poorly. Take “alive.” We *notice* things that are alive, and know how to describe them. Of course, many more things are *not* alive than alive. Astonishingly, in describing *them*, we fumble. What is a rock? “Not alive,” “inanimate,” and “lifeless” say what it *isn't*, not what it *is*. “Dead” implies “once alive.” We seem to have a blind spot: either something is (or was) alive, or it isn't ... anything.

If we have only two basic words—“alive” and “dead”—we imply a clear distinction between the two. But is it so simple? Is “alive” or “dead” like “white” or “black?” A sharp line between them? No gradual transitions?

We don't understand how not-alive first became alive. So far as we know, it happened only once on earth, about four billion years ago. Somehow, the mixture of molecules on the prebiotic earth—earth *before* life—taught itself to live. It probably wasn't a single, sharply defined event—*not-alive to alive*!—but what it was we just don't know. Now, life always comes from life: cells divide, fish spawn, babies are born.

We understand the transition from alive to not-alive much better, since we, ourselves, die. We also know that death sometimes seems gradual; but when, exactly, does an animal or a cell cease to live? How do we describe it? Perhaps we need a continuum between “alive” and “not-alive”? Alive; partly-alive; dead but with some chemistry still running; *really* dead?

A virus walks the line. Outside the cell, it's just a package of biological molecules: DNA or RNA, and a capsule that contains them. It is no more alive than other small clumps of molecules. When it encounters an attractive cell, it sticks, and forces its DNA inside. This viral DNA enslaves the cell: the molecular machinery the cell normally uses to replicate itself—to make more life—is forced instead to make more virus. So, outside the cell, the virus is dead. Inside the cell, the virus becomes an integral part of something living.

What then is a virus? Not-alive? Potentially alive? Occasionally part-of-something-alive? Not-alive but able to control life? Something else? If we accept that there are things—viruses, desiccated spores, frozen embryos—that are neither clearly alive nor dead, but somewhere in between, the world becomes complicated.

We know we have ethical obligations to living things. But what are our obligations to things that are not fully alive, or are ambiguously not-alive? To think about life and near-life in fresh ways, we need new concepts and new words.

WHY CARE?

Knowing how things work is a great pleasure to some, uninteresting to others. Curiosity is the itch that understanding scratches.

Not everyone itches; but everyone wants drinkable water. "Understanding" makes the water drinkable, and the trains run on time.

It's fascinating to understand how electrons perambulate through the forests of atoms in a silicon crystal. It's satisfying to puzzle out the design of the millions of switches we call the Internet. It's calming to phone Grandma in Ulaanbaatar and wish her Happy Birthday!

为何在意?

有些人对事物运行的来龙去脉感到津津有味,有些人则会觉得乏味至极。

好奇心就像是皮痒时对挠痒的渴望,并不是每个人皮肤都会发痒,当然亦不可能每个人都会对科学抱有强烈的好奇心。但是,人人都得喝水,总得要有人去弄明白科学的那点儿事,我们才能喝上纯净的水,坐上准点的火车。

其实,在科学的世界里徜徉,尤其是进入到"小的世界"中,还真会让我们欣赏到不一样的风景。弄明白电子是如何在硅晶体浩瀚的原子海洋中漫游会让人着迷;苦苦思索而弄明白如何在因特网中设计成千上万个开关会让人满足;同样,远在乌兰巴托时能给奶奶打电话祝福她生日快乐会让人安心。

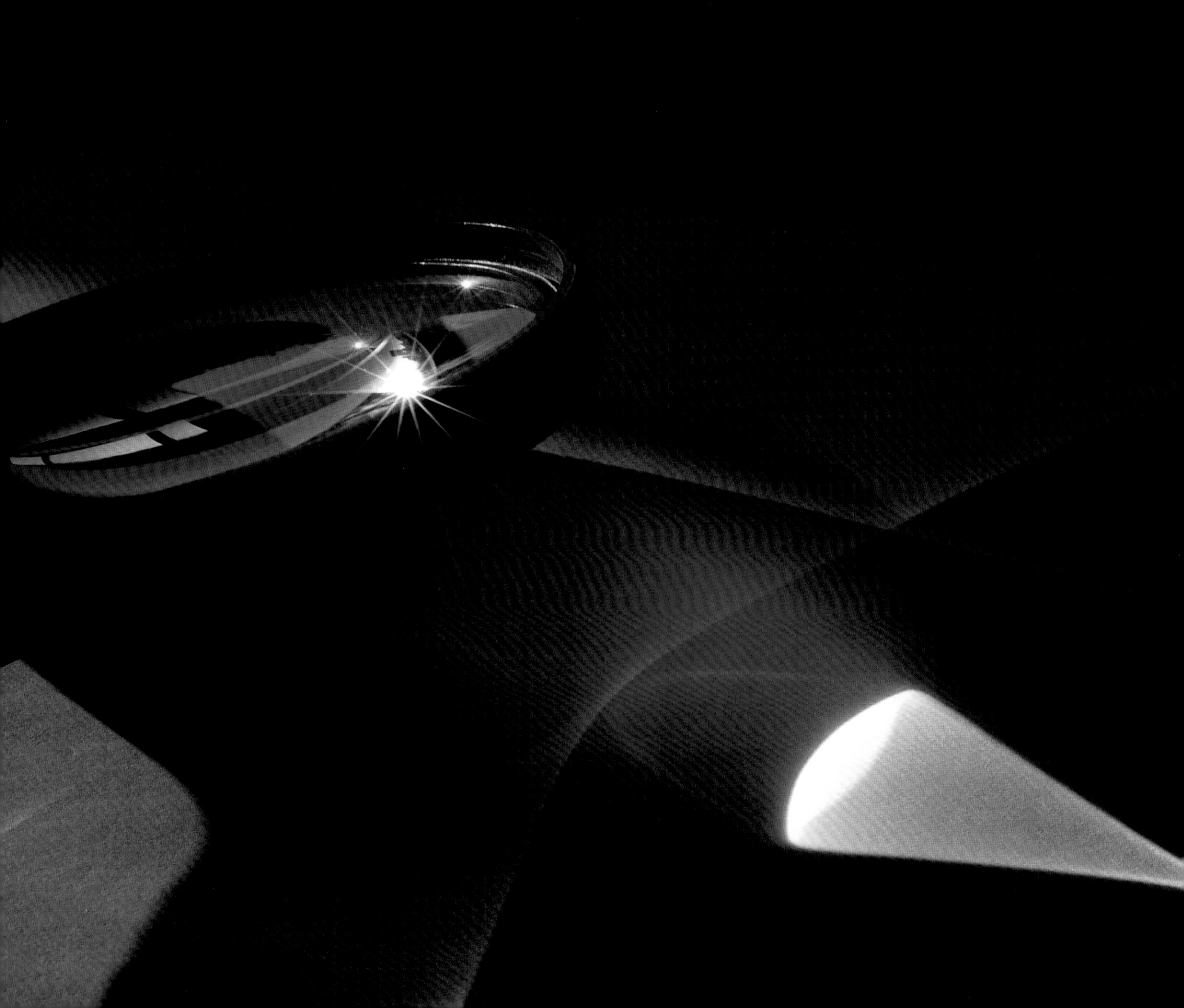

光　刻

Writing with Light

31

提起纵火，虽然不是件好事，但确是孩提时的一大乐趣。

在家长眼里，火无异于恶魔。为了让孩子们免遭这个恶魔的侵扰，大人们总会把火柴锁起来。但对于小孩来说，从父母身边偷出火柴不过是小菜一碟。另外，点火的方法有多种，除了火柴这一现成的工具，孩子们还能很快学会利用透镜和太阳光来点火。在阳光明媚的夏日，他们拿着透镜对着太阳，就能将光聚焦成一个炙热耀眼的斑点，把干枯的树叶放在斑点，那么几秒钟内便会冒烟，并在一分钟内燃烧。这个点火的过程十分简单，只要拿着透镜的手不动即可。

如今，同样的方法也可用来制作含有复杂线路和晶体管的微电子芯片。这种技术被称之为“光刻”，即用光来勾画图案。通常用工业化的激光器作为光源，这种光虽然肉眼无法可见，但其实比太阳光还强。当光斑被聚焦到一层薄如肥皂泡的光敏有机膜时，就能写出一个完整的、具有特定形状和功能的微电子系统。

更进一步，如果我们把有机膜沉积在打磨过的硅片上，并再铺上一层镂空的不透明蜡纸，这样光打上去后就能在镂空的地方镌刻。这种方法能大面积地同步制作多个细微结构且速度极快，看起来有点像传统的黑白印刷术，但它的分辨率要高很多。经过进一步加工，这些光刻出来的图案便成为芯片，进而成为各种数字化的器件，如手机、智能手机、Wii 游戏机等的“大脑”。就像被不断纺出的糖丝层层包裹的棉花糖一样，这些不断更新的器件在我们的生活中占据越来越重要的位置。

ONE OF THE JOYS OF CHILDHOOD is setting fires. Too bad, but it happens.

There are many ways to start a fire: the simplest is with stolen matches, though wise parents lock matches away to frustrate the demons who periodically hijack their children. Children quickly learn that a lens and sunlight work almost as well as matches. On a bright summer day, a dry leaf can be smoking in seconds, and flaming in a minute. To focus the light of the sun into a brilliant, and *hot*, spot requires only a child's steady hand.

The same strategy forms the intricate traceries of wires and transistors that become microelectronic chips. *Photolithography* is the name of this technique—writing patterns with light. It starts with the brilliant flare of an industrial laser—brighter than the sun, but invisible to our eyes. Moving a focused spot of this light across a film of photosensitive organic matter thinner than the film of a soap bubble draws the patterns that give integrated circuits their shapes and functions.

A faster method of writing patterns shines the light through an opaque stencil onto a polished disk of silicon coated with the photosensitive film, and forms many lines at once. The process is like black and white photography, but with enormously higher resolution. The images so produced become, with manipulations that are probably among the most clever that humankind has ever produced, the nanoscopically structured brains of the devices—cellphone, smart phone, Wii—that spin the cocoons of digital cotton candy in which we spend much of our lives.

Eleanor Rigby

《埃莉诺·里格比》

32

MUSIC DOES SOMETHING WE DON'T begin to understand: it transmutes sound into emotion. The Brahms *Requiem* brings us exaltation; Muzak, boredom; the Doors, desolation.

"Eleanor Rigby" is a song by the Beatles: it generated a sense of isolation and failed opportunity, and reflected, perhaps, the incompatible mixture of hope and resignation that permeated society during the nuclear terrors of the '60s. We all wanted to hear "Eleanor Rigby." Since the Beatles could not sing to everyone, we used machines to sing to us all.

We have invented many ways to replicate music. With learning and practice, we ourselves can sing; with records, magnetic tapes, CDs, iPods, and disk drives we can be sung to. This image magnifies the epitome of an almost forgotten but much beloved form of recording. It shows several tracks of a 33-rpm vinyl recording of "Eleanor Rigby." The delicate needle of the record player rested in the grooves; as the disk turned, the needle followed the wiggles imprinted there. The tiny vibrations of the needle—side to side and up and down—generated voltages in a cleverly responsive crystal; amplification of these voltages provided power to speakers. Our ears transcribed the resulting waves of pressure in the air into waves of pressure in the cochlear fluid, analyzed them by tone, intensity, and duration, and shipped the data to our brain. The brain generated a perception that we call music, and an emotion we call despair.

From the nasal whine of the Beatles, through oscillations in plastic, air, and fluid, to emotion. Remarkable.

音乐的功效常常出乎我们的预料——它能将声音转变成我们的情感,比如勃拉姆斯的《安魂曲》让我们意气风发;酒吧里的背景音乐让我们百无聊赖;大门乐队演唱的摇滚乐让我们颓废和忧伤。

《埃莉诺·里格比》是披头士摇滚乐队演唱的一首歌。这首歌所带来的孤独和挫败的感觉也许正反映了20世纪60年代人们对弥漫在社会中的核恐怖既祈盼又无奈的矛盾心情。一代又一代的人们都爱听《埃莉诺·里格比》,但披头士却不可能永生并为每一个人演唱。庆幸的是,我们有了复制音乐的装置。这样,即使在没有披头士的现今,我们也能跨越时空欣赏这首歌。

人类已经发明多种复制音乐的方法。最初只是自己学着唱,后来用唱片、磁带、光盘、ipods以及计算机硬盘把音乐拷录下来,然后再播放。这张图片展示的是曾经被人们所喜爱但现已被人们淡忘的黑胶唱片。这个图案正是刻录在唱片上的《埃莉诺·里格比》的一部分。当唱片转动时,探针便会随着表面的图案而不断跳动。在播放唱片的整个过程中,探针的上下或左右的微小移动都会在响应的晶体里产生电压,这些电压进一步放大后便可作为电流提供给喇叭,成为不同的声音。我们的耳朵可以先将空气中的压力波转化为耳蜗流体中的压力波,然后分析它们的音调、强度和持续时间,最后将这些信息传送到大脑。于是,大脑就产生了一种我们叫做"音乐"的感知和随之而来的"绝望"情绪。

披头士的歌声被封存到黑胶唱片中,再经空气传播直至在耳蜗流体中振荡,最终让人产生情感,这一过程是多么的神奇啊!

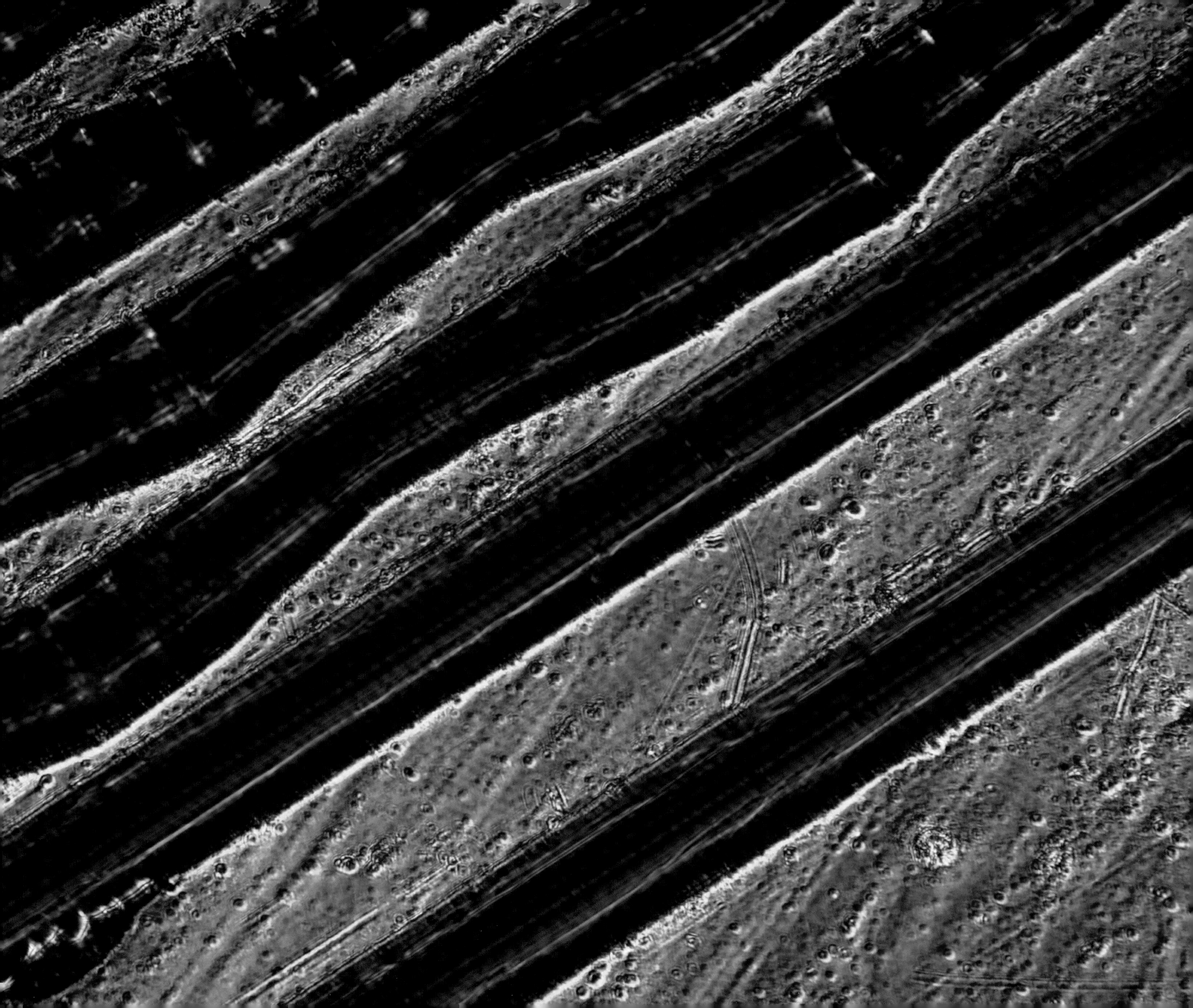

算　盘

Abacus

33

我们不知道人类在多久以前就会计数了。考古学家们发现两万年前的骨头上有着一道道的划痕，这说明那时人们就开始用一套比现在的1,2,3…更为复杂的数字系统来计数。当我们想要知道一群羊的头数或者多少片面包能换到一只羊时，计数就不可避免。很多时候，我们采用扳手指的方法，这也就是十进制的由来。

但是用手指来计数容易出现差错，而且并不是每个人的手指都那么灵巧，也不是每个人都有完整的十根手指。因此，我们的祖先也曾尝试采用不同的方法和工具来辅助计数，例如在棍子上刻痕、把绳结在一起、在罐子里放鹅卵石等，直到某一天他们发明了算盘。算盘已经在人类的发展史上使用了上千年，它的结构和工作原理异常简单：下层的五颗珠子表示从1到5，而上层的两颗珠子每一颗都表示5。

既然有了手指和算盘，计算机为何不用类似的方法来计数呢？首先，十根手指并非是天意。假如进化的过程中发生变异的话，我们也可能只剩三根手指。当然，这样一来，我们就必须准备另一套钢琴曲目，同时我们也会自动认为三进制才是最合理的。更为重要的是，与我们使用太多根手指常常会出现差错一样，如果使用太多的数字，计算机也会容易出错。而我们如果只使用0和1这两个数字（即二进制）来计数，就能最大限度地避免出现差错和歧义，也比较容易纠正错误。

It's not clear when humans began counting. As long as 20,000 years ago, scratches on bones suggest a system of numbers more sophisticated than "1, 2, many ..." When we needed to know the number of sheep in a herd, or perhaps how many loaves of bread bought one sheep, we began to take counting seriously. With 10 fingers, a number system based on 10 was an obvious choice.

It's easy to lose track with fingers. Also, not everyone's fingers are equally agile, nor does everyone have 10 fingers. We tried many different systems of numbers, and devices, to help in counting: notches on sticks, strings tied together, pebbles in pots. The abacus is so simple that it has survived for millennia. The five beads on a bottom wire count 1 to 5; the two beads on the top wire each stand for 5.

Why not use an equivalent system—one based on the natural arrangement of 10 fingers, or on the abacus—for counting with computers? For one thing, there is nothing natural about 10 fingers; with a slightly different throw of the evolutionary dice, we would have three fingers, an entirely different piano repertoire, and the instinct that "natural" was powers of 3. More importantly, in the same way that it's easy to get confused with lots of fingers, it's easy to confuse a computer with too many kinds of numbers. The least confusion, the least ambiguity, and the easiest checking for errors come when there are only two numbers: 0 and 1—or "binary."

Counting on Two Fingers 二进制计数

34

In counting, almost anything works: raised fingers, pebbles in a pile, knots on a string. We gravitate toward 1, 2, 3, 4, 5, 6, 7, 8, 9, 10 because evolution gave us ten fingers. Computers don't fret about evolution: they do what they are told.

Why is binary best for computers? With only two possibilities, it's difficult to make mistakes. Any system with only two unmistakably different states will do: 0, 1; red, blue; full, empty. The rules of binary addition are also simple: 0 + 0 = 0; 0 + 1 = 1; 1 + 0 = 1; 1 + 1 = 0 (and carry 1).

These are glasses viewed from above; the red liquid is a good Shiraz. Think of pouring liquids from glass to glass, and you understand how a computer adds. A 0 is an empty glass; a 1 is a filled one. The glasses show the addition of two binary numbers: 001100 + 010110 = 100010. (On our fingers, this addition would be 12 + 22 = 34.) A computer, adding these same numbers, would pour electrons from one container (one we call a capacitor) to another. In binary, counting with wine and counting with electrons works equally well.

In their electronic world, computers can be distracted by noise. When they err, we suffer: GPS takes us to a cornfield rather than a restaurant, or the telephone drops a crucial word: "I ... you!" Noise—the overheard whispers from other electronic conversations; the bursts of static from an occasional cosmic ray or a witless shark gnawing on an undersea cable; the low level of crackle on the line that physics tells us is always there—all cause mistakes. No system is perfect, but binary resists the confusions of noise better than any other system of counting. For computers—which make sharks seem like towering intellects—simplest is best.

几乎所有的东西都可以用来计数，比如扳手指、将鹅卵石分堆、在绳子上打结等。我们之所以倾向于用十进制来计数是因为人类在进化过程中产生了十根手指。相比之下，计算机就自由多了，我们告诉它怎么做，它就会怎么做。

为什么二进制的计数系统最适合计算机呢？主要原因在于这样的方式在处理数据时只存在两种可能性，要么是 0 要么是 1、要么黑要么白、要么满要么空，这就大大降低了出错的几率；而且二进制的加法规则异常简单，只有四种情况：0+0=0，0+1=1，1+0=1，1+1=0（并进 1）。

图中所示的是从上往下看到的玻璃杯，红色的液体是一种上等的 Shiraz 葡萄酒。试想一下怎样将这些液体从一个杯子转移到另一个杯子，你就会明白计算机是如何工作的了。"0"代表一个空杯子，那么"1"就代表一个装满葡萄酒的杯子。图中的这些杯子向我们展示了如下二进制数字的加法：001 100 + 010 110 = 100 010。如果我们用扳手指的办法来进行这个运算将会得到 12 + 22 = 34。一台计算机把这些相同的数字加起来时，其实是把电子从一个容器转移到另一个容器，这种容器通常称为电容器。对于二进制来说，用葡萄酒来计数或用电子来计数并无任何差别。

在电子的世界，噪声会干扰计算机的工作。计算机一旦出错，我们就会遇到麻烦，例如全球定位系统可能带我们去一片玉米地而不是我们想去的餐馆，或者我们打电话时可能会被略掉一个关键字，诸如"我……你！"之类的。那么，噪声是什么呢？它是其他电子设备之间"交谈"时传来的杂音，是偶然的宇宙射线爆发所产生的静电干扰，是因为愚蠢的鲨鱼在啃海底的电缆，是物理学上讲到的通讯线路中总是存在的噼噼啪啪的响声……这些都会导致计算机出错。没有什么系统是完美无瑕的，但是二进制系统在抗噪方面表现得比其他任何计数系统都要出色。对计算机而言，最简单的计数系统便是最好的系统。

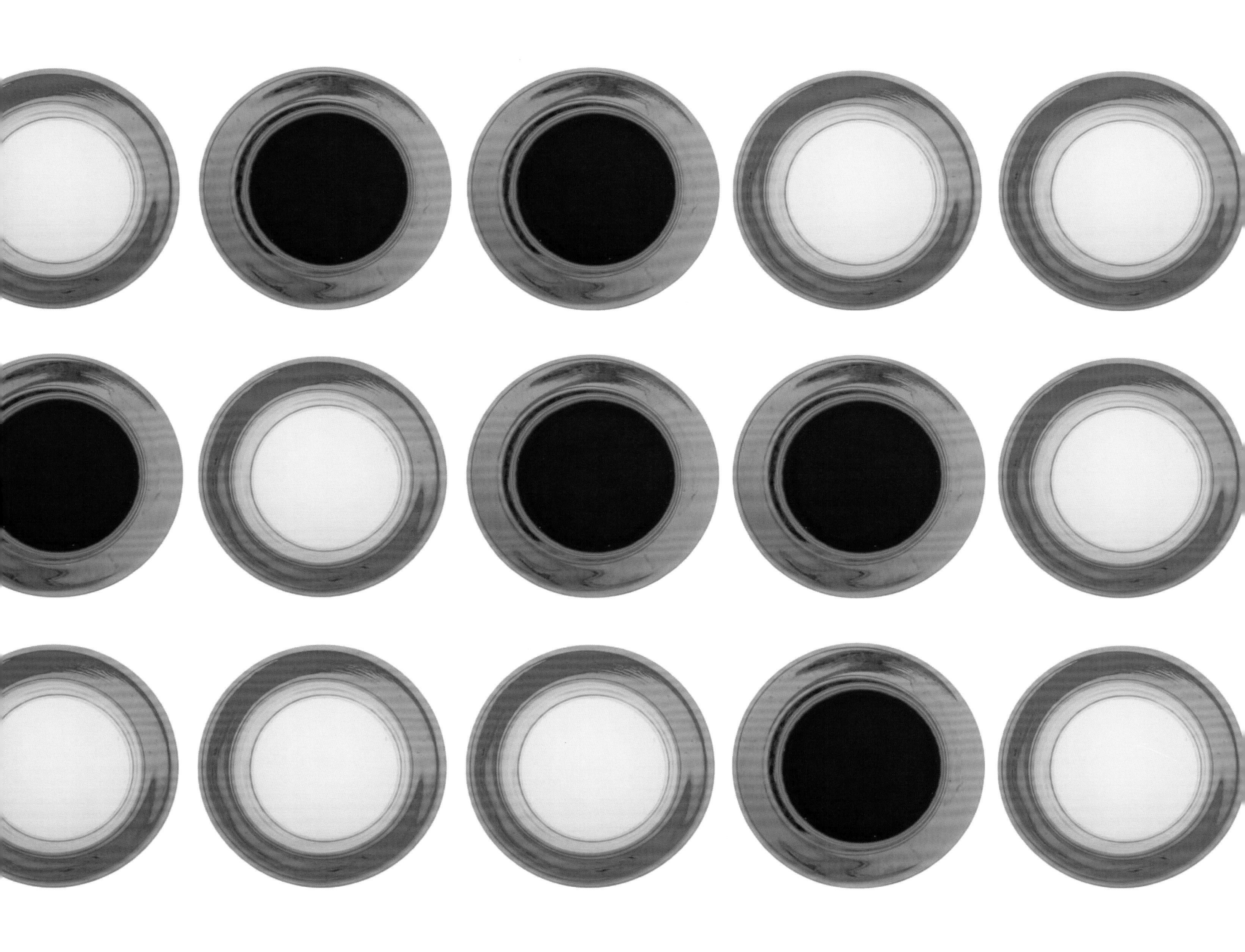

PORTION OF BABBAGE'S DIFFERENCE ENGINE.

巴贝奇的计算器

Babbage's Computing Engine

35

为了操控物体，人类发明了数之不尽的机器。然而，有这么一部机器，像拉斯科洞窟壁画、单音律的圣歌、莱特兄弟的飞机一样值得称颂，它的诞生标志着一个新时代的开始。这部机器是由英国人查理·巴贝奇在1849年发明的，它可以用来进行数学运算，是世界上最早的可编程计算机。

巴贝奇的机器与当今的电子计算机几乎没有任何共同之处，但它提供了最初的“针线”，使得我们后来可以“编织”出万维网。巴贝奇对测量和数字十分着迷，他甚至已经认识到了“信息”这个在当时不为人知的概念，但他肯定没有料到光纤及计算机网络会在某一天覆盖全球。即便如此，他的初衷仍然是正确的。他的机器是为了提供一个可靠的算术平台，从而为一些不专心或容易出错的人在进行加减法运算时分忧。他还试图将这种机器广泛地装配到其他机器中来辅助运算。在当时，巴贝奇的机器确实具有十分广阔的应用前景，但从未真正被制造出来。不过即使制造出来，那个家伙估计会重达数吨。

巴贝奇的机器就是一台巨大的机械加法机，它与现今随处可见的计算机微处理器（包括手机中的那种）的不同之处，就好比一个滴水的水龙头与尼亚加拉大瀑布之间的差异。一方面在于它们的元件大小不同，巴贝奇的机器使用直径为厘米大小的机械和齿轮，而计算机芯片上的线宽度却小于20纳米，这样的宽度仅相当于50个金原子紧密排成的线长度。更为重要的是它们之间不可同日而语的运算速度，装配微处理器的计算机的速度比巴贝奇理想中的机器的运算速度要快上万亿倍。

TO ADMIRE THIS MACHINE is to admire the cave paintings at Lascaux, plainsong, or the Wright brothers' airplane: it was a beginning. We humans have invented countless machines to manipulate objects. This machine—designed by the Englishman Charles Babbage in 1849—was invented to manipulate numbers. It would have been the first programmable computer.

Babbage's engine has almost nothing in common with electronic computers, but it wove the first thread in the filigree we call the world-wide web. Babbage was obsessed with measurement and numbers. Although he would have recognized the word "information," he could not have imagined a planetary-scale lacework of optical fiber and computers. Still, he had the right idea: his engine was intended to solve arbitrary problems in arithmetic, and thus to lift the burden of addition and subtraction from the stooped shoulders of distractible, error-prone humans and place it on the broad shoulders of machines. Babbage's engine had very broad shoulders indeed. It was never completed, but had it been, it would have weighed several tons.

The difference between Babbage's engine—a large, mechanical adding machine—and the computing engines (microprocessors) everywhere, even in cellphones, is the difference between a dripping faucet and Niagara Falls. Part of the difference is the size of the components. Babbage used machined wheels and cogs centimeters across; computer chips use wires less than 20 nanometers wide—a width spanned by 50 atoms of gold. Speed is even more important: microprocessors compute perhaps a trillion times more rapidly than the machines Babbage imagined.

Computers as Waterworks

供水系统般的计算机

36

WHAT, REALLY, IS A COMPUTER? Think of a municipal water supply. It's a system of pipes and valves that pumps water from reservoir to kitchen. The pipes are hollow iron tubes, and the valves are ... well ... valves. A computer is not so different. It's a system of pipes and valves that pumps streams of electrons rather than streams of water from place to place. Of course, nothing is hollow: the pipes and valves for electrons are made of solid metal and silicon (usually with a soupçon of added boron or phosphorus). Still, the electrons flow through silicon more smoothly than water through a cast-iron pipe.

But it's not where the electrons go that counts, it's how they *get* there, and how valves-for-electrons (called transistors) control their flow. With proper valving, the fluid—the electrons—can do binary arithmetic. Because the transistors—the valves—open and close quickly, this arithmetic-carried-out-by-pumping-electrons-between-containers is extraordinarily rapid (currently, about a thousand billion additions per second).

Binary numbers can also be used to represent almost anything else you like—letters, colors, pitch, intensity of sound—by assigning identifying labels. The assignments are arbitrary—the letter a has the binary name 01000001, and the punctuation mark*!* is 00100001—but are universally readable, since everyone subscribes to the same conventions. From these assignments grow word-processing programs, text messaging, and smart phone images.

And that's all there is: selling a load of fish by cellphone in Thailand, writing a besotted love letter on a laptop in Indianapolis, downloading Beethoven's Ninth in Tierra del Fuego, landing a 747 after a long flight from Mumbai—it's all just moving and recognizing and adding incomprehensibly large numbers of 1s and 0s incomprehensibly rapidly.

Technological reductionism does take some of the romance out of writing love letters in Indianapolis. Maybe we should stick to longhand for some things.

计算机是怎样工作的？试想一下城市的供水系统，它是由众多管道和阀门交织而成的网络将水库里的水源源不断地输送到千家万户的厨房。管道是中空的铁管，阀门则负责管道间的开通和关闭。计算机与供水系统没有太大的区别，它也是由管道和阀门组成的。不同的是，计算机依赖的是电子输送，而不是水。当然，在计算机系统里没有什么东西是中空的，传输电子的管道和阀门是由硅(通常会添加痕量的硼或磷)和固体金属制作而成。此外，与水在铁管中流动相比，电子在固体元件中的流动会更快更稳。

不同的是，计算机不会对电子进行计量收费，它关注的只是电子如何到达目的地，以及如何被我们称之为晶体管的阀门来控制。当受特定的阀门控制时，电子可以进行二进制的运算。由于晶体管能快速地开和关，电子传输的速度也就快如闪电，目前的运算速度已达到每秒上万亿次加减。

二进制的数字可以用来表达像字母、颜色、声调、音量等几乎所有的东西，而这一切都只需给它们分配一个可以被识别的代码，例如字母“a”在二进制里被分配的代码是“01 000 001”，标点符号“！”的代码则是“00 100 001”。虽然代码的分配是任意的，但只要我们每个人都遵守这些约定俗成，这些代码便是谁都可以读懂的。基于这些代码，我们便有了文字处理程序、发送的短消息，以及智能手机中的图片。

在泰国通过手机卖鱼，在印第安纳波利斯用笔记本电脑写一封让人沉醉的情书，在火地岛下载贝多芬的第九交响曲，一架从遥远的孟买飞来的波音 747 飞机正在降落……所有这一切都是由大量的 0 与 1 通过快速转移、识别与组合来完成的。

技术的发展会使我们在印第安纳波利斯写情书时少一点浪漫，也许我们应该返璞归真，还是用笔来写些东西。

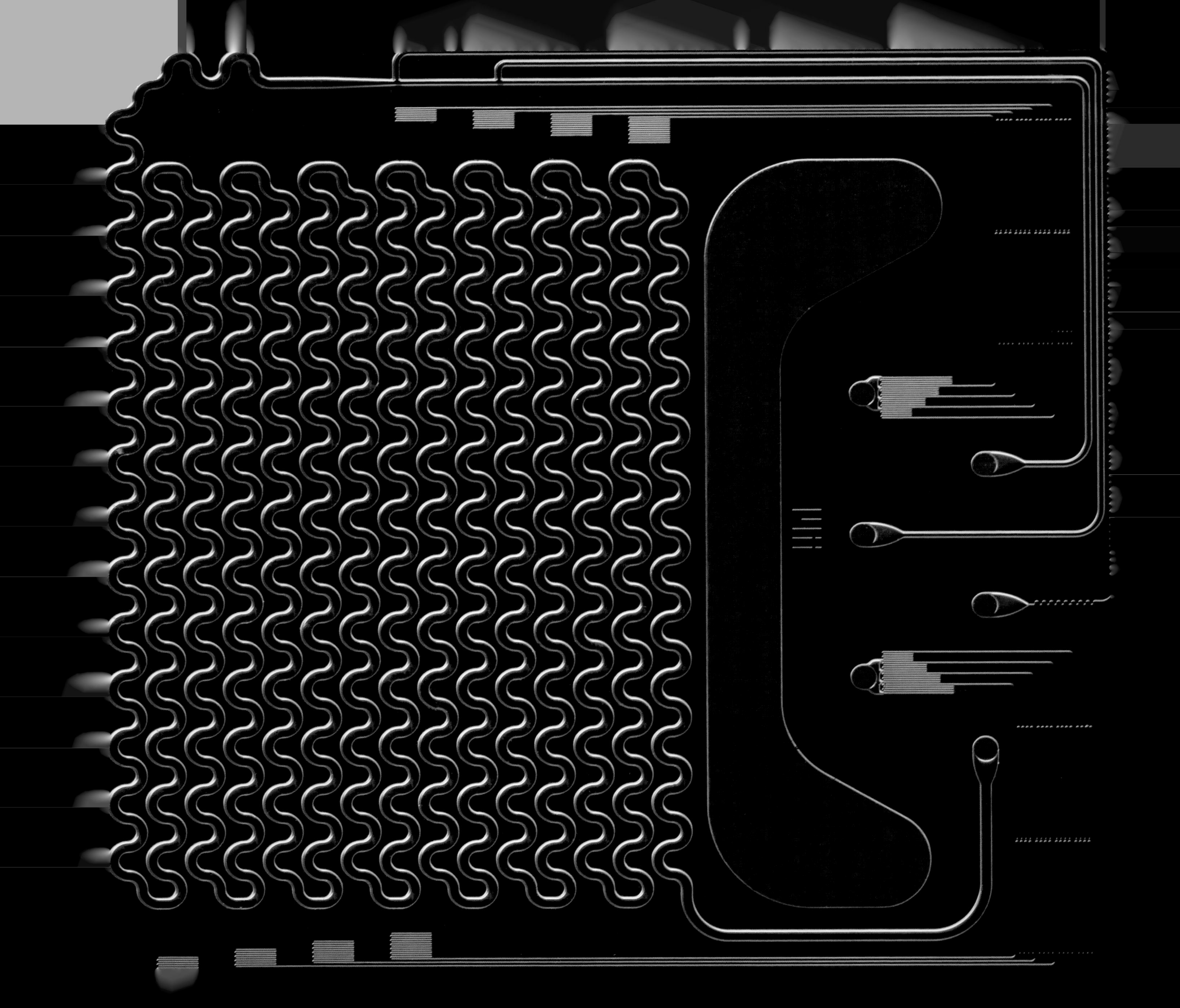

微反应器

Microreactor

37

几乎所有的制造过程都充满危险。一辆成品汽车看起来挺扎实的，但它的制造过程却是危机四伏。例如，它是用极高温度的液态钢铸成的，冲压时的巨大压力足以在1毫秒内把我们压成肉饼，稍不留神激光焊接机就能把我们的身体切成两半……如果但丁明白制造业对我们的人身安全有这么大威胁的话，他在写《地狱》时就会有更多的灵感。也正因为如此，专家们一直在不懈努力试图使工厂里的操作尽可能的安全。

有些东西生来就是安全的，比如软床垫或棉花糖就不大可能伤害到谁。还有一些东西由专业人员操作要比业余人士危险小得多。比如厨房工具(如火柴)在一个厨师手里很安全，但在她六岁的儿子手里则另当别论。当然，有些东西则生来就是很危险的，比如高温的物体、有毒的化学品、爆炸物等等。我们该怎样来使用它们呢？

对待那些在工业制造过程中必须使用而毒性又很强的化学药品，一个有效的办法就是每次只制备非常少的量，同时尽量让它们生产后就消耗掉。这样的策略可以将事态恶化的危险降到最低。微反应器就是一种能实现这个目标的装置，通过在芯片上设计一系列的通道，能快速持续地操作危险的化学反应，并且总是以微量的形式进行。

MAKING ALMOST ANYTHING can be dangerous. An automobile seems solid enough, but it comes from blazing cauldrons of liquid steel, stamping presses able to squash us into sticky films in a millisecond, and laser welders that would not notice if they cut us in half. Had he understood the indifference of manufacturing technology to human comfort and safety, Dante could only have been inspired. Still, managed by experts, industrial processes do operate safely, or at least safely enough.

Some things are intrinsically safe: it is difficult to come to harm with mattresses or marshmallows. Others pose little risk in the hands of experts but are dangerous to amateurs: kitchen matches are safe in the hands of a cook, but not in those of her six-year-old son. Some things are never entirely without danger: objects at high temperatures, toxic chemicals, explosives. How should we handle them?

One approach to dealing with the necessary and incidentally toxic chemicals occasionally used in technology is to make them only in small quantities, and consume them as rapidly as they are produced. That strategy minimizes damage if something goes awry. This device is a microreactor—a set of channels on a chip designed to carry out dangerous chemistry quickly, continuously, and always in very small quantities.

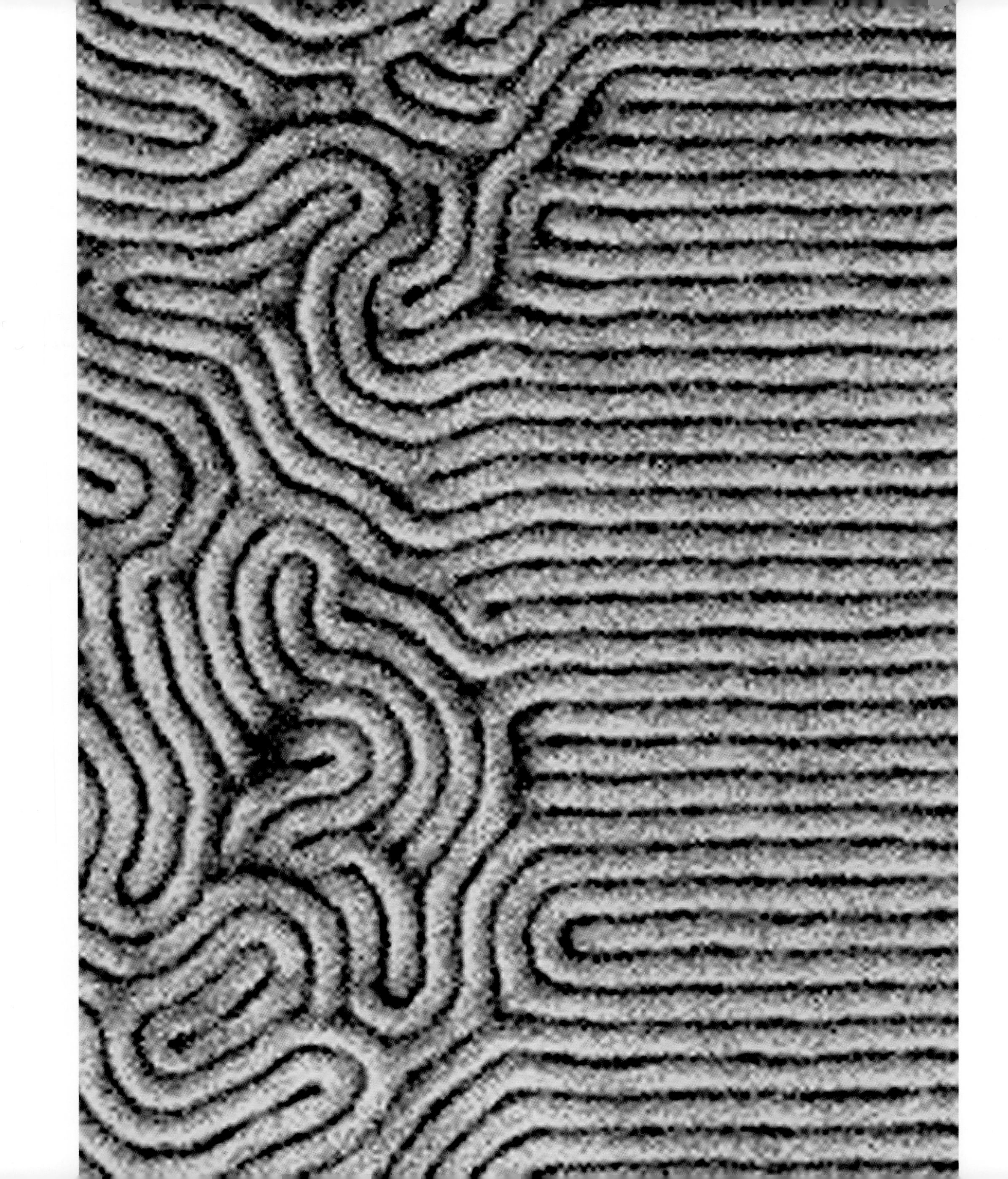

模板效应

Templating

38

假设你需要将分子或纳米级的胶体粒子组装成一个有序的结构，比如晶体或一个电子产品的某些元件，你会怎么做？怎样才能让你的手指小到能够把如此微小的东西放到合适的位置上呢?这当然是做不到的。最好的办法莫过于你能使它们按照你想要的方式进行自组装；同时，你亦可以用一个模板或图案来引导它们，让它们紧紧跟着你的想法走。

试想一下，如果要让一群人在下页图中所示的椅上坐下来，你就明白什么是模板效应了。这些椅子朝着一个方向摆放得整整齐齐，这样一来，不管是在一场婚礼、毕业典礼、还是葬礼中，它们都可以使杂乱的人群很快地安顿下来。当每个人都坐下时，一幅有序的图案也就形成了。在这一过程中，椅子充当了模板的角色。

那么，分子是否也有自己的“椅子”呢？它们的“椅子”又是怎么样的呢？直到现在我们还不知道最佳的答案，但仍然在努力尝试去揭示这个问题的真谛。晶体生长过程中的晶面或许可以作为纳米级组装的模板，分子在晶体表面形成有序的阵列，新到的分子可以在原有分子上排起来，一层又一层地进行自组装。

左页的图片展示了在晶体表面形成的分子阵列，所用材料是一种叫做聚合物的大分子。这些分子有两个不同的末端，在自组装时，相同的末端会聚集在一起，形成几十个纳米宽的长行。在图片的右半部分，晶体表面刻有平行条纹的模板，因而分子被引导排列成了平行有序的队列；而另外半边，由于没用模板，分子则变得杂乱无章。

有时候粗糙的模板也会给人带来意想不到的惊喜。例如在制作液晶显示器时(现在绝大多数的手机或电脑都使用这种显示器)，我们需要铅笔形状的液晶分子有序地堆积在一起，因其有序程度低于晶体但高于液体，故被称作“液晶”。在显示板表面轻轻地用布擦一下会有助于液晶的形成，但我们仍然不是十分清楚这个过程是怎样工作的。也许摩擦的过程中会产生微米级的划痕，对于液晶分子来说，在其上进行组装可能类似于让一群人站在一块犁过的田里。犁沟可以帮他们排成一定的图案，却不能像成排折叠椅那样让他们排列得更为有序。

Suppose you need to assemble molecules or nanoscale colloids into a regular assembly—a crystal, say, or some part of an electronic device. How would you do it? How could you ever develop fingers small enough to put such tiny things into their place? You couldn't. The best you could do is to *encourage* them to assemble spontaneously as you would like them to, and to give them instructions—a template or pattern they could follow.

Imagine a large group of people, and the chairs on the next page, and you'll get the idea. The chairs are ordered; they are regularly spaced; they point in one direction. They invite, command, orient a milling crowd—to sit down for a wedding, a graduation, a funeral. As everyone takes a seat, the crowd assembles itself into a pattern set by the template.

"Chairs" for molecules? What would they be? We don't know the best answer and are still testing strategies. The face of a growing crystal is a familiar template for nanoscale assembly. The molecules exposed at the face form an ordered array: new molecules joining the crystal sit on top of the ones already in place, so the crystal grows, layer by identical layer.

The image opposite shows rows of molecules (*large* molecules, called polymers) on a surface. These molecules have two dissimilar ends, and the "like" ends cluster together into long rows a few tens of nanometers wide. In half the image, the molecules are guided into a regular pattern of parallel rows by a template of parallel stripes written on the surface. In the other half, there is no template, and the rows of molecules wander, unguided.

Sometimes a crude template works surprisingly well. The operation of liquid crystal displays (the displays on most cell phones and computers) requires pencil-shaped molecules to stack in a fashion that is *partially* ordered (less-ordered than a crystal, more ordered than a liquid; hence, "liquid crystal"). Gentle mechanical rubbing of the surface of the display in contact with the liquid crystal seems to guide this ordering. It probably produces microscopic scratches, but we don't know exactly how the process works. It might be similar to asking a crowd to stand in a tilled field: the parallel furrows would provide a pattern of sorts, but it would not produce the order of folding chairs.

催化剂颗粒

Catalyst Particles

39

在出色的厨师手里，即便是牛奶、面粉、糖、胡萝卜这样简单而普通的食品原材料也可以变成一桌丰盛的佳肴。

催化剂是一种能使别的物质发生变化而自身保持不变的物质，它的作用宛如一名厨师。在厨房里，厨师可以将普通的食材神奇般地变为人们难以忘怀的美味，但是当烹饪结束、食客赞许之后，他（她）仍然还是厨师，并未发生任何改变，还将一如既往地准备下一次的烹饪。

这张图片向我们展示了纳米级的金属颗粒。这些间距约为1纳米的线条代表着一层层的金属原子，它们是透射电子显微镜聚焦的电子束（只有几纳米粗细，就像从手电筒中发出的光，但相比之下十分昂贵）在投射过程中与颗粒中的金属原子作用而形成的。采用这种金属颗粒（例如钴和铂的合金）来制备的燃料电池可以更加有效地利用氢气和氧气来产生电能。一些类似的颗粒（更大量并附载在粘土和陶瓷上），还可以将原油转变为汽油和飞机燃料，或者加工用于制药的分子。当化学反应结束以及反应产物被清除之后，这些纳米颗粒依然还在那里而没有发生任何改变，它们就是所谓的催化剂。

一种好的催化剂，宛如一名出色的厨师，能用普普通通的食材制备出弥足珍贵的佳肴。而差的催化剂，就如同一名拙劣的厨师，即使使用了上等的食材，烹饪出的食物也是无人问津。

MILK, FLOUR, SUGAR, CARROTS—in the hands of a chef, groceries become a feast.

A "catalyst" is something that causes a transformation but is not changed in the process. A chef transforms ordinary ingredients into a sensory delight; but after the cooking is done and the meal is exclaimed over, she is still a chef, and ready to cook again.

This image shows nanoscale particles of metal. The lines, separated by about 1 nanometer, are layers of atoms (or, rather, the *shadows* of these layers, cast in the light of a *very* expensive "flashlight" that focuses a beam of electrons, rather than light, into an illuminated point a few nanometers in diameter). This kind of particle (a mixture of platinum and cobalt atoms) is designed to make fuel cells extract electricity more efficiently from hydrogen and oxygen. Similar particles (in multitudes, supported on particles of clay and ceramic) transform crude oil into gasoline and jet fuel, while others process the molecules used to make drugs. After the chemistry is over and the products have been removed, the catalyst is still there, unchanged.

A good catalyst, like a good cook, starts with inexpensive ingredients and makes something valuable. A bad one, like a bad cook, makes a pig's breakfast.

Christmas-Tree Mixer

圣诞树般的混合器

40

SMALL VOLUMES OF LIQUID can be important. Only a drop of blood is available from a finger prick, and only a drop of urine from a newborn human or an adult mouse. If you have a few rare cells, you must keep them in small volumes of liquid just to avoid losing them. Dangerous chemicals are always best used in small quantities. There are many reasons to be interested in small volumes.

Microfluidic systems make it easy to handle very small volumes of liquid, but mixing is a problem. To mix two liquids—even viscous ones—in large quantities, you use a motor, and attach it to a propeller, or a screw, or rollers: think egg beater or dough mixer. Microfluidic systems do not accommodate motor-driven mixers: it's difficult to attach motors to hair-sized channels, or to insert propellers into them. An alternative is to force the liquids to flow as thin, adjacent streams, and then let the diffusion of molecules between streams do the mixing.

Mixing is also useful in forming gradients. A *gradient* is a continuous change in something with position. For example, the annoyance caused by the neighbor's car alarm going off decreases with distance: up close, it's unendurable; the farther away you get, the less awful it is. This device is a microfluidic mixer that makes gradients in biochemicals. Two fluids—red and blue—flow from the top into a network of 100-micron-scale channels. They converge, mix, divide, reconverge, remix, and redivide, forming a smooth gradient from red to purple to blue at the bottom. This kind of mixer—using biological signaling molecules rather than dyes—allows cell biologists to ask a cell living in the exit channel a question: Do you like a higher concentration of what I'm offering you, or a lower one? The answer to that question helps biologists understand how cells know what organ to form during embryonic growth, and how they find an infected cut on a leg.

很少量的液体对我们来说也可能十分重要,例如扎破手指取出的一滴血,新生婴儿或成年老鼠排出的一滴尿。如果你有少量罕见的细胞,你必须将它们保存在小体积的液体中以免找不到。对于危险的化学药品来说,减少使用量也是降低风险的最佳办法。所以,我们有太多的理由去关注微量的体积。

微流系统使我们能很容易地去操控体积很小的液体,但是如何使如此微量的液体有效混合,却是个大难题。当大量混合两种液体时,即使是非常的黏稠,你也可以用马达驱动的搅拌器来加速混合,就像打蛋器或和面机一样。微流系统却不适合使用马达驱动的混合器,因为要将马达与头发丝粗细的通道相连接,或是在这么细微的通道里插入螺旋桨确实是太难了!另一种方法是让液体以薄的平行层方式来流动,然后通过界面处分子的扩散来进行混合。

混合也是形成浓度梯度的一种有效方法。梯度表达的是某个参数随位置的连续变化。例如,邻居的汽车报警器给我们带来的烦恼主要取决于我们跟车之间的距离。靠得越近,就越让人无法忍受;距离远一点,由它带来的烦恼也就少一点。图片中展示的这个装置是用来产生生物化学分子浓度梯度的微流体混合器,红色和蓝色两股液体从顶端出发流进 100 微米大小的网状通道中。这些液体在通道中汇聚、混合、分开、再汇聚、再混合、再分开……这样反反复复数轮过后,最终在底部形成了由红色到紫色再到蓝色的一系列渐变色的梯度。如果用生物信号分子代替五颜六色的染料,这样的混合器便可以产生这些分子的浓度梯度,从而让细胞生物学家对身处通道中的细胞提出这样的问题:你喜欢我提供给你的高浓度环境呢,还是低浓度的? 这个问题的答案将有助于生物学家弄明白在胚胎发育阶段细胞是怎样知道应该发育成什么样的器官,以及细胞是如何在腿上找到被感染的伤口的。

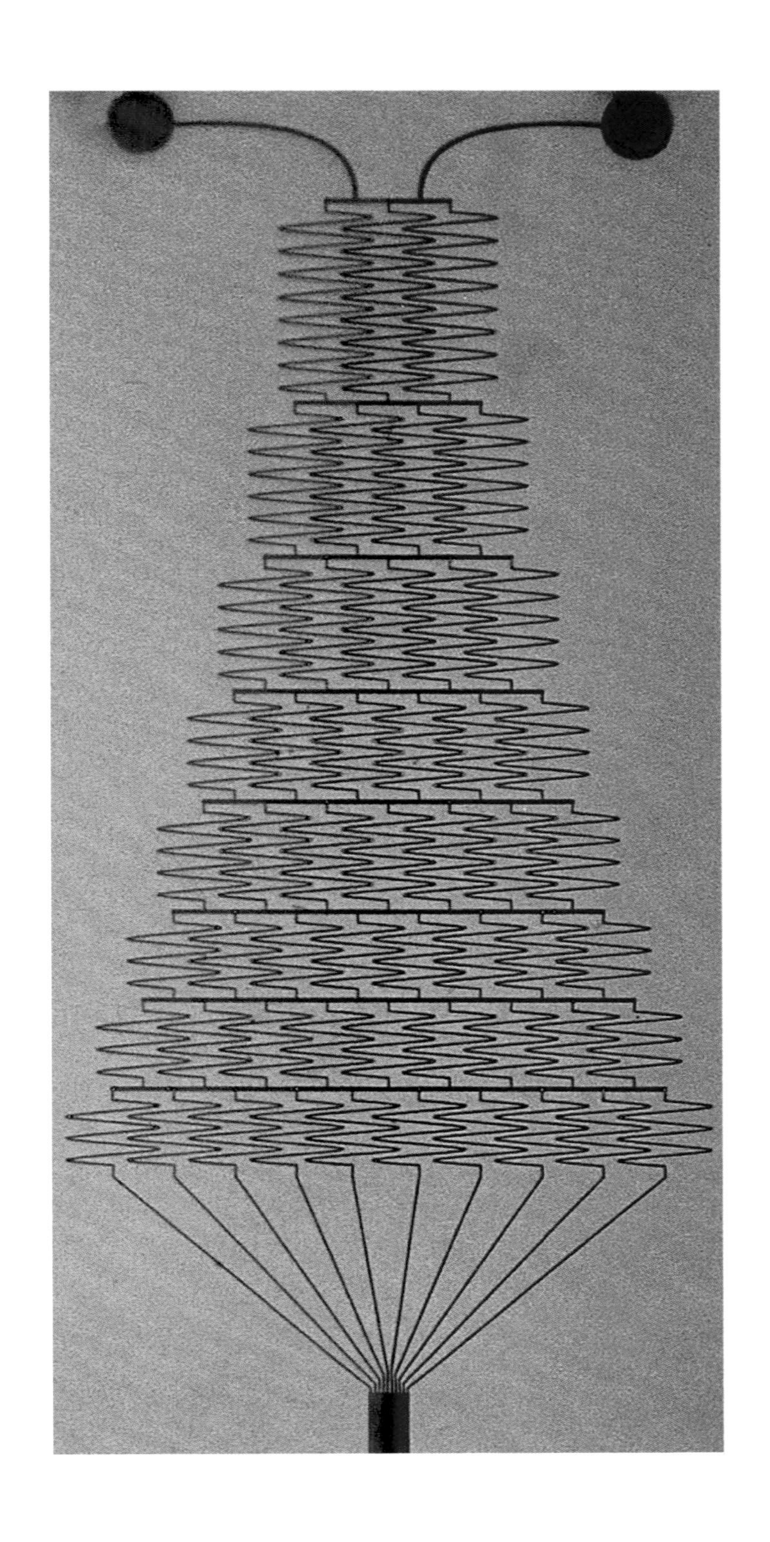

自组装

Self-Assembly

41

我们的世界里充斥着各种机械，它们能为我们做一些人类自己做不到的事情：飞机让我们自由翱翔；手机让我们在超越耳语的距离闲谈；Wii 这种电视游戏机使我们不必为找一位网球搭档而烦恼……我们利用机械进行加工、精制、浇铸、钻孔、焊接、栓接、冲压、铸模、粘合以及移植。我们还制造出结构复杂的机器人来完成更复杂的工作。

我们做事可以很有效率，但往往会缺乏条理。有解决这个问题的途径没有？当然有，生命为我们提供了最令人惊讶的例子。我们的身体并非是通过分子一个接一个的方式胡乱拼成的，相反地，包括大分子、细胞、组织和器官在内的身体各级零部件都是按照合适的次序自发排列和组合在一起的，即自组装。这种组装完全是自发性的，并不需要人或外界机械的协助。

除了生物体以外，自组装广泛存在于世界万物之中，只是我们平时没注意而已。大气中无序的水分子在低温下形成冰晶，冰晶又能自发排列成为结构复杂却又精致的雪花；盐田中，随着海水的缓慢蒸发，盐渐渐析出形成炫目的多面体；沙粒在风的作用下形成密集的沙丘；开瓶后的香槟在表面形成由微小气泡组成的有序团簇……

这张图片展示的是放置在两个底部轻微凹陷的盘子里的厘米大小的玻璃球。不需要人为干预，只需轻轻拍打几下盘子的边缘，这些玻璃小球便会整齐地“结晶”成非常有序的阵列。就这么简单！

试想，如果我们把计算机零件放在一个桶里，轻轻地摇动几下，能否自动组装出一台崭新的电脑呢？这样的好事固然不会在明天发生，然而，哪怕是朝着自组装方向迈出的一小步，也将为我们节省很多力气。

We fill the world with machines that do for us what we can't do for ourselves: airplanes let us fly; cell phones let us gossip when we are beyond whispering range; Wii makes it unnecessary to find a tennis partner. We machine, refine, cast, drill, weld, bolt, stamp, mold, glue, and implant. We build robots—themselves complex machines—to help us.

We are effective but sometimes inelegant in making things. Is there another way? We *know* there is: life provides the most astonishing examples. Cells, and organisms comprising many cells, assemble themselves. We do not fabricate a baby, molecule by molecule. Her pieces—molecules, cells, tissues, organs—put themselves together in the proper order: they *self-assemble*, rather than requiring assembly by external devices or by people.

Many forms of self-assembly are familiar, if often unnoticed. The disordered molecules of liquid water order themselves into crystalline ice; ice spontaneously grows into intricately shaped snowflakes. Salt forms beautiful polyhedra when sea-water evaporates slowly. Sand pushed by the wind forms dunes. The tiny bubbles on the surface of champagne organize themselves into clusters before they burst and their spray tickles your nose.

This picture shows centimeter-sized glass spheres resting in two dishes with slightly concave bottoms. With a few taps on the side of the dish, they "crystallize" into a pleasingly regular array. We can admire, but we don't have to be involved. It's easier that way.

Will we ever be able to put the parts of a computer into a bucket, shake gently, and see a computer spontaneously emerge? Not soon; but even small steps in the direction of self-assembly could save enormous effort by humans and machines.

Synthetic Nose

人造狗鼻子

42

WE AND DOGS—OUR SPECIES' ANCIENT COMPANIONS—sense the world in complementary ways. We see; they smell. Perhaps that's why we've gotten along: we hunt well together, and we protect and amuse one another.

Protection! Our lives depend on it. Dogs could smell the lion we could not see. And although lions are now not an everyday problem, we rely on dogs to discriminate odors—molecules drifting in the air—where we see nothing. They sense strangers at night; fires while we sleep; drugs hidden in suitcases; the tracks of lost children. They can also detect nitrated organic compounds wafting from explosives.

Although dogs are unparalleled in their combination of sophisticated nose and substantial intelligence—and *very* skilled at detecting bombs—we would prefer to save them for safer jobs. And, of course, dogs can be fallible at a job where a mistake can be fatal: they have their moods, exhaustions, and periods of ennui.

To relieve dogs of some of the burdens and hazards of being our "noses," we are trying to mimic their discriminating powers of smell. This sequence of images shows the response of an "artificial nose" to a puff of odor. Each spot is a dot of polymer—an organic material we normally encounter in sandwich wrap, disposable coffee cups, and glue—containing a dye that gives off visible light when illuminated with ultraviolet light. The emitted color is very sensitive to the presence of other molecules dissolved in the polymer. When an odor in air drifts across the assembly of spots, the odor molecules dissolve in the polymers, and their colors shift; when the molecules evaporate, the colors shift back. These patterns of drifting colors can sometimes answer the crucial question: "Explosive or sausage?"

Such systems are primitive compared to a dog's nose, but for detection of a land mine, primitive may be enough. And it might allow us, for once, to honor our obligation to protect our gifted friends.

人和狗自古以来就是密不可分的伙伴，并一直以互补的方式感知这个世界——我们看，它们闻。也许这正是我们相处甚欢的原因，我们一起打猎、相互保护和互相娱乐。

离开保护，我们将无法生存。狗能保护我们是因为它们能嗅到我们看不到的狮子，尽管狮子已不再是我们生活中每天都面临的威胁。狗还能通过气味来帮助我们辨别漂浮在空气中的分子。狗能在夜间第一时间感知陌生人的到来，在我们沉睡的时候发现火灾，找出藏在行李箱中的毒品，追踪到遗失的小孩；它们还能发觉易爆物品中硝基化有机物散发出的特殊气味……

狗的嗅觉与判断能力确实是无与伦比的，这使得它们在检测炸弹时游刃有余。当然，它们也可能在一些工作中会犯下致命的错误，因为它们毕竟是动物，也有自己的情绪，也有筋疲力尽的时候以及倦怠期。如果在探测地雷时犯错，那将会是致命的！因此，我们有时候会于心不忍，总想让它们干些更安全的工作。

为了消除这些情绪化影响带来的不确定性，也为了使人类和狗都免于不必要的伤害，我们试图模拟狗鼻这种借助嗅觉来辨识事物的能力。右图中这一系列的图片展示了一种"人造狗鼻子"对气味的响应。我们所看到的圆点是一种含有染料的聚合物，这种聚合物是我们日常生活中最常见的一种有机材料，广泛应用于三明治的外包装、一次性咖啡杯，以及胶水中。当我们用紫外光去照射这种聚合物时，它里面的染料就会发射出可见光。而光的颜色对存在于聚合物中的其他分子十分敏感。当空气中的气味飘过这些圆点时，气味的分子会渗透到聚合物中，使得这些圆点的发光颜色改变；而当这些分子挥发掉以后，这些圆点发出来的光又回归到原来的色彩。当含有不同染料的圆点组成一个阵列时，它们发光的图案可以用来辨别气味，告诉我们这个气味是来自炸药还是腊肠。

与狗的鼻子相比，这个系统实在太简陋了。但如果仅仅用来检测地雷，这样简陋的系统还是足以应付的，至少可以让我们对身边这些天赋异禀的狗类朋友尽到应有的保护责任。

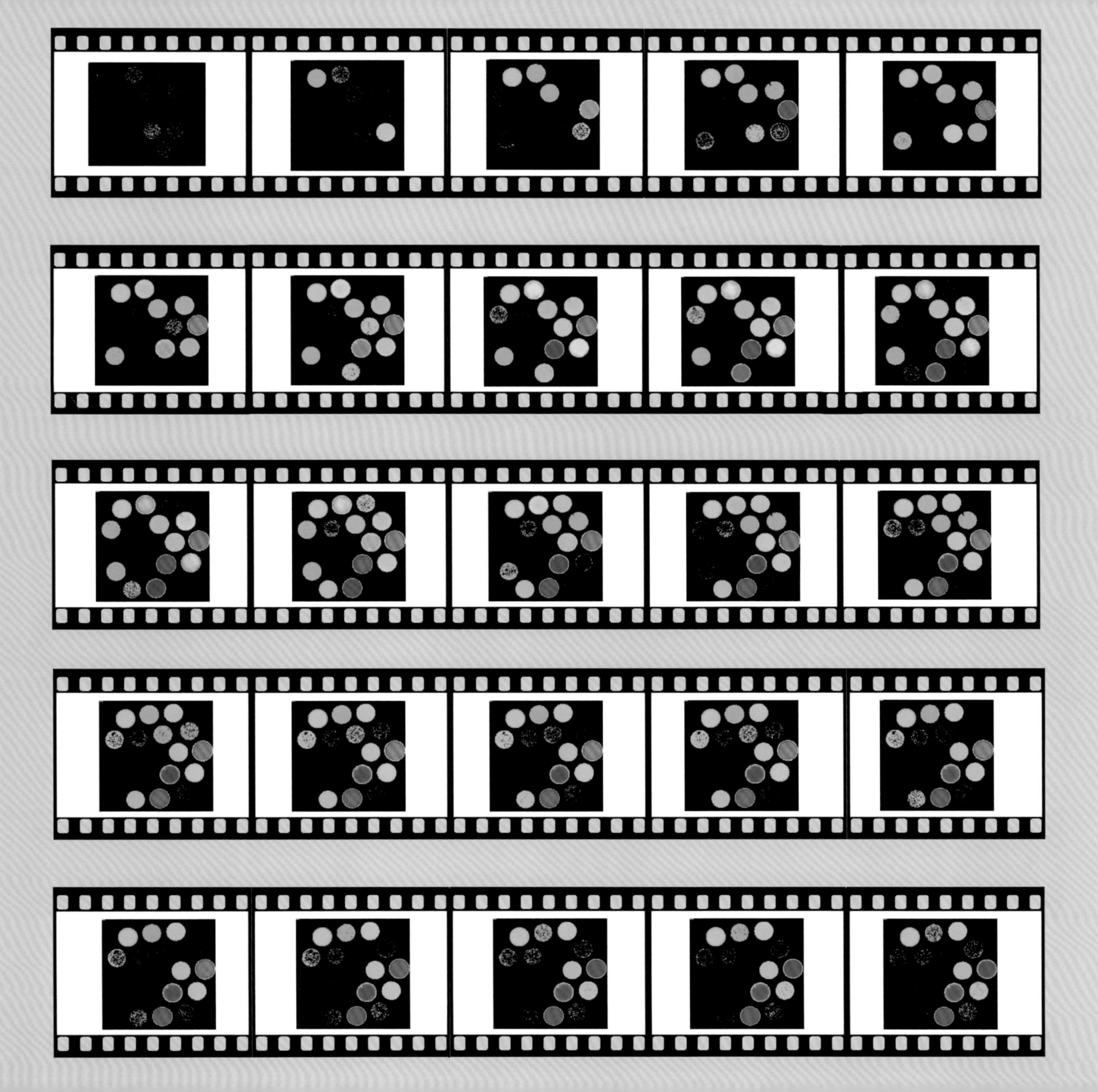

千 足 虫 Millipede

43

在数字化的世界里，"信息"不是从智慧、慈祥的祖母嘴里讲出来的故事，而是由 0 和 1 组成的冷冰冰的字符串，它从一个地方流向另一个地方，或者只是默默地滞留在某个地方。

因为庞大的信息量，我们必须能够快速地读取这些"比特"（二进制单位）并将其储存在尽可能小的空间里。在传统的微电子器件中，"比特"反映的是掺杂硅中某块用于容纳电子的一个小区域，或者是压入塑料光盘表面并可用激光读取的凹坑。用更小的凹坑来记录信息，是实现计算机更高密度存储的新方法。然而，怎么才能快速地读出这些信息呢？

在轻微加热的情况下，扫描探针显微镜的探针提供了一种有效读取信息的途径。这个探针与原子力显微镜的探针十分相似，当它非常接近一个快速旋转的表面上比较平坦的部分时，就会把热量传递到这个表面；如果探针在小凹坑的上面，探针与旋转表面距离拉大的同时热量传递的过程便会轻微地减慢。当储存盘在探针下面高速旋转时，探针尖端的温度会随路径相应地发生迅速而微小的变化，从而把储存在盘表面的信息读取出来。

仅仅使用一个探针达不到快速读取信息的目的。但多个探针同时工作的话，还是可以的。图中所展示的这种装置就像一条千足虫，它有着上千根手指般的探针，通过协同作业，便可以实现快速读取信息的目的。

"INFORMATION" IN OUR digital world is not stories from the mouths of wise, warm grandmothers, but cool strings of 1s and 0s streaming from place to place, or resting in storage.

Because we use bits (*bi*nary digi*ts*) in enormous numbers, we must be able to read them quickly, and stow them in the smallest possible space. In conventional microelectronic devices, bits are either small regions of doped silicon engineered to hold electrons, or pits pressed into the surfaces of plastic disks and designed to be read with lasers. One new approach to computer memory with high densities would also record information as pits, but smaller ones. But how to read them?

A warm, scanning probe microscope tip would provide one way. If the tip—very similar to the tip of an atomic force microscope—is close to a flat part of the surface spinning rapidly beneath it, it loses heat to it; if it is above a pit—that is, momentarily further from the surface—it loses heat slightly more slowly. The pattern of tiny but very rapid changes in temperature of the tip as the disk spins beneath it provides a record of the pits.

One tip cannot read information rapidly enough to be useful. A thousand tips can, if they read the disk at the same time. The millipede, the device shown here, is a step in this direction: rather than the thousand feet of its namesake, this device has a thousand fingers—a thousand separate tips—and can read a thousand patterns of pits at once.

E-paper and the Book

电子纸和电子书

44

FOR CENTURIES, we've entrusted our memories, our records, our musings, our imaginings to books. Many centuries, many people, many memories. That's a lot of books. Too many to schlep around; too many to search for the one essential fact; too many to mine for understanding or insight or wisdom, or for entertainment and escape and flights to other worlds.

The Book (as booklovers ironically refer to the object fabricated in paper and ink) now also competes with the world-wide web. The web makes all data—details of everything, imaginative triumphs of language and thought, facts, factoids, falsehoods, drivel—democratically available to anyone.

And where does this leave The Book? The answer is: in limbo. Books have advantages. Their printed words last longer than words held in the magnetic cobwebs of computer memories. They don't use electricity. And people who have lived with books like their heft, smell, and feel, and the fact that you can slip them under mugs of hot coffee to avoid damaging the table.

The electronic book—the e-book—is a shotgun wedding of the aristocratic book and the vulgar web. The electronics of the e-book are straightforward; the display is the difficult part. The "page" of this e-book is a sheet of plastic covered with many tiny electrodes. The sheet includes droplets of oil containing a mixture of black and white nanoparticles. When an electrode has one charge, the black particles go to the back of the sheet and white light reflects from the front; when that electrode has the opposite charge, the color changes from white to black. Simple electronics stores the original text of the book and converts images of its pages into patterns of charge on electrodes, both using very little power.

The e-book may be the salvation of writing, or the death of it, or it may be an evolutionary eddy—like the platypus—in the co-evolution of humans and their stories. It is as convenient as the book—maybe more convenient. It needs batteries, but what doesn't now? It doesn't quite look like a book, but it holds well over a thousand complete books, and weighs almost nothing. It doesn't smell like a book, and turning a page is a different experience, but for people who have never been hooked on the smell of books, or the whisper of a turning page, perhaps neither counts as a great loss.

"Call me Ishmael" is the opening line from a beloved nineteenth-century paper book. Three letters are replicated here on the plastic page of a twenty-first century e-book.

自从有文字开始，人类就不断地将自己的经历和对世界的看法记录在纸上。太过久远的年代，太多的人和太多的记忆成就了浩瀚的书海。且不说携带这些书是多么的费劲，在其间寻找某一关键的细节又是多么的困难，过多的书往往还会左右我们的智慧、阻碍我们对现实的洞察与理解，更有甚者往往会在书中迷失自我，沉湎于其中而不可自拔。

即便书（那种由纸张和油墨制作成的东西）的影响力如此之大，现如今也不得不与万维网相互竞争。网络把所有的数据，包括各种各样的细节、文字游戏中虚构的胜利、事实、所谓的事实、谎话、蠢话等等毫无保留地呈现给每一个人。这样一来，纸质的书将何去何从呢？

答案是：可能被遗忘。事实上，纸质书还是有很多优点的。印在纸上的文字要比存放在电脑磁盘中的保留得更为长久，并且也无需耗费电。而且对于那些爱书如命的人来说，纸的手感、油墨的气味以及那种无法言语的感觉足以让他们陶醉。此外，你甚至还可以将书垫在热的咖啡杯下以免烫伤桌子。

相比之下，电子书则是高雅书籍和粗俗网络的任意结合。电子书的制作是比较简单的，而如何显示则相对困难一些。电子书的页面其实是由很多微小电极覆盖的塑料膜所组成，在两个电极之间，又有许多细小的油滴，其中含有黑色和白色的纳米颗粒。当电极带上特定电荷时，黑色的纳米粒子就会跑到塑料片的底面，白光便从顶面反射回来，整个页面呈现出白色；当电极带上相反的电荷时，颜色又会从白变为黑。如此简单的操作可以用来存储文字信息，同时还能将页面上的图像转换为电极上的电荷分布。所有这些操作都不需要消耗太多的电能。

电子书或许是写作的救星，抑或是它的终结者，当然也可能是一个进化的特例，就像鸭嘴兽一样，在与人类同时进化的过程中演绎着他们自己特殊的故事。电子书不会比纸质书复杂，甚至还更简单。所不同的是，它需要电源，但现在又有什么东西能离得开电源呢？虽然看起来不像书，但它却涵盖了上千本书籍的完整信息，并且一点重量也没有。它虽然闻起来不像书，翻页的感觉也完全不同，但是对于那些从来没有沉迷于纸香味或翻页声的人来说，这根本算不上什么大损失。

"Call me Ishmael"是我甚为喜爱的一本19世纪的纸质书的卷首语，其中的三个字母刚好显示在一页21世纪的电子纸上。

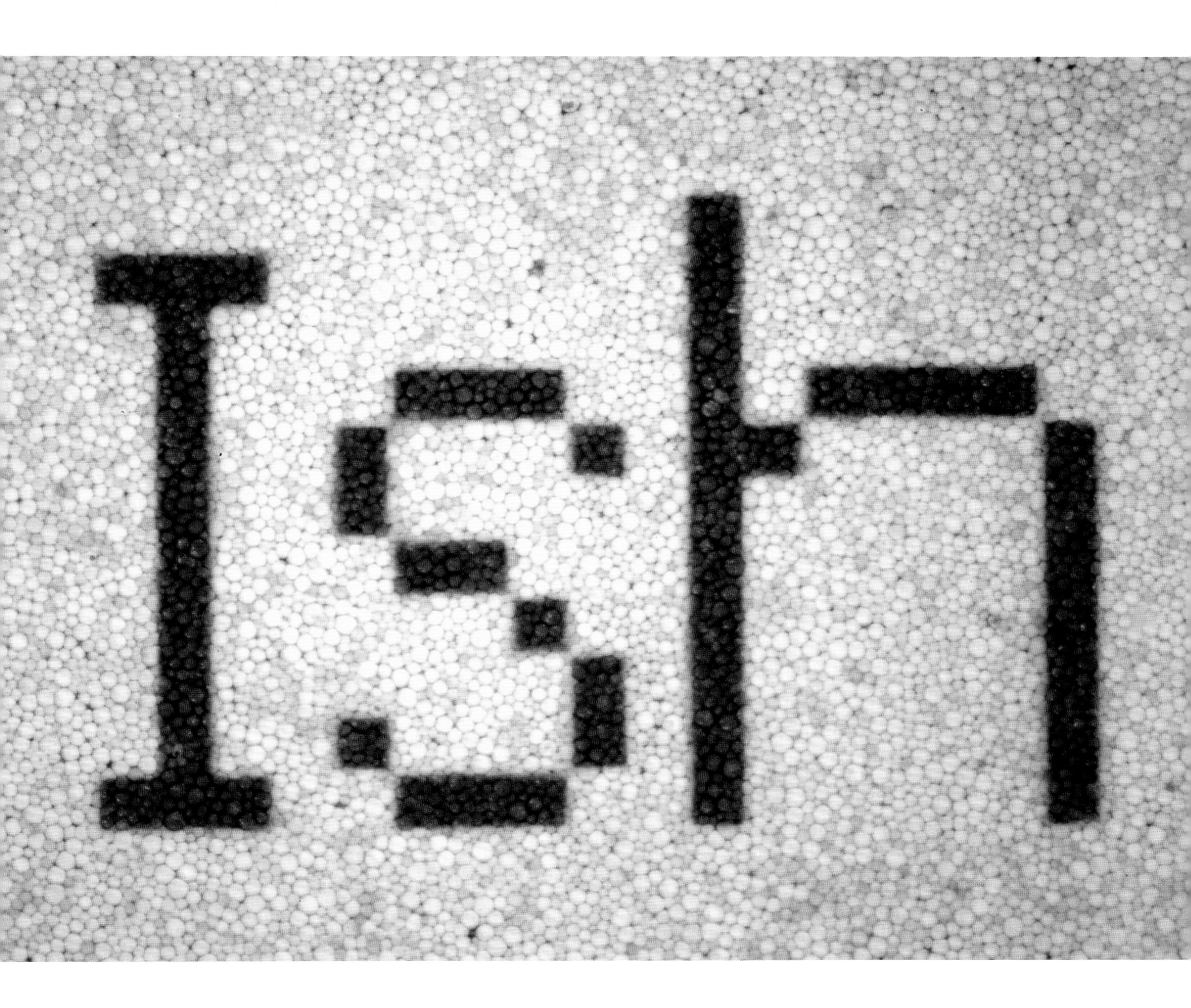

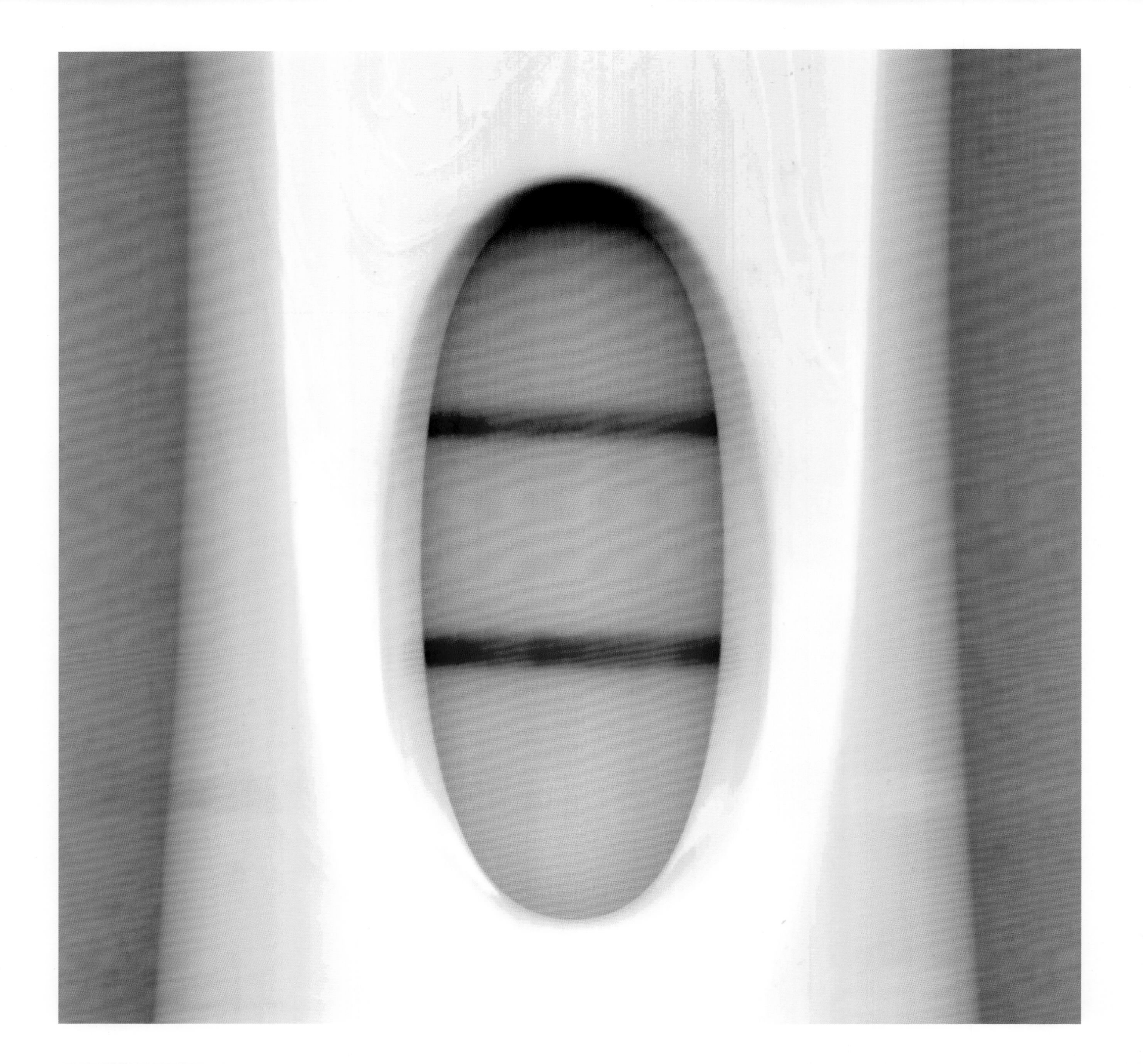

决定命运的横向流动检测法

Lateral-Flow Assay as Crystal Ball

45

这是一个再普通不过的情景：一个不再年轻的女人买了一支怀孕测试棒，想看看运气如何。这种测试棒是将看不见的小分子打印在一条纸片上，然后将其装进一个塑料的支架中。当尿液通过塑料套管末端的引流纱布条进入测试棒，便能告诉她是否已经怀孕以及接下来该怎么做。尿液会在试纸条上从一端向另一端扩散，就如同咖啡滴在餐巾纸上的情形一样。如果尿液中含有绒毛膜促性腺激素(英文缩写为hCG，即怀孕早期产生的一种激素)，这些分子就会吸附在从试纸条上释放出来的红色纳米粒子上并一起扩散至另一区域。当它们遇到含有另一种与hCG有强烈作用的分子组成的线时，就会被吸引而最终停留并聚集在那里，并使这条线显现出红色纳米颗粒的颜色。如果检测时试纸条上显示的只有一条红色的线，那便证明尿液中没有hCG分子；如果有两条红线的话，则表示有喜了。

这样的测试原理与一个游戏极其相似：如果你是Sue(人名)，你就捡起一个红色的沙滩球并站在那里，于是我们就能知道手里拿着红色沙滩球的那个人就是Sue。怀孕测试与这个游戏相比，除了名字是hCG而非Sue，其他的都一样。如果说还有什么不一样的话，那就是怀孕测试可不是一个游戏。

一条细细的红线就能辨别一个女人怀孕与否，这能让她知道上天是青睐于她而送给她一个孩子还是再次忽略了她。

AN OFT-REPEATED FAMILY DRAMA:

The woman, not quite as young as she once was, buys a pregnancy test—a strip of paper printed invisibly with small patterns of molecules and encased in a plastic holder—urinates on the wick at the end of the plastic casing, and gazes on her future. The urine wicks its way along the paper like coffee across a napkin. If the urine contains molecules of hCG—human chorionic gonadotropin (a hormone that appears shortly after pregnancy begins)—these molecules attach to red nanoparticles released by the paper, and the two move together in the fluid to another region where they are drawn down in a line by a second molecule printed in the paper. One line means only “the test is working”; two lines mean “you’re pregnant.”

It’s like a game: If your name is Sue, pick up a red beach ball, and go stand over there so we can see who you are. Except in this game, the name is hCG, not Sue, and it is not a game at all.

One thin red line marks the difference between pregnant and not-pregnant, and tells whether Fate will favor the woman with a child, or has again ignored her, at least this time around.

Testing Drugs in Cells

用细胞来筛选药物

46

WE HAVE A TOUCHING BELIEF THAT with enough magic we will live almost forever. Enough drugs, enough exercise, enough chickens tied in forests, and age will scuttle away. Exercise *does* help, but it's tiring. What about drugs?

Drugs are complicated. Some work, some don't; some work sometimes, in some people, with some illnesses. All have side effects. Developing a new drug is horrendously difficult: somewhere among the millions of kinds of molecules in you—the patient—a drug must locate and stick to just a specific few, and divert them from whatever mischief they are causing.

Drugs are molecules that do good, while doing as little harm as possible. They must also be profitable. (For the time being, our society has decided that "health" is a profit opportunity, not a social obligation.) Drug developers identify candidate molecules and audition them first in animals—mice, rats, hamsters, dogs. Successful auditionees are introduced into humans cautiously, in the highly regulated, slow, and expensive process known as "clinical trials." Most fail. The central technical problem is that humans are not—however similar they might in some ways seem—large, furless mice. Although different species of mammals have similarities in metabolism, there are also critical differences, and discarded molecules that cure disease in mice, but not in people, litter the laboratories of pharmaceutical companies.

How should one test molecules for effectiveness and safety in humans, if testing is almost prohibitively slow, expensive, and ethically and legally constrained? Testing in human *cells*—available from tissue removed in operations, or separated from blood—offers one new approach. True, human cells are not intact humans, and the difference between a cell and an organism is large. But human cells contain the entire human genome and have the essential features of human biochemistry.

There are many, many ways of looking at a cell. Optical microscopy generated this image showing internal structure, painted with differently colored dyes. A cell biologist, looking at this structure, can gauge a cell's *joie de vivre*, and infer how it responded after tasting a bit of a molecule whose creators aspire that it grow up to be a drug.

我们都有一个美好的愿望，那就是能够长生不老。的确，如果我们有各种各样的药物，又坚持每天锻炼身体，吃原生态的食品，似乎就应该能使青春永驻。锻炼的确会有所帮助，但它也会让我们的身体感到疲倦。那五花八门的药物又会怎么样呢？

药物是非常复杂的，当我们生病时，某些药物能起作用，而另一些却一点用也没有。还有一些药物只能在特定时候起作用、对一部分人起作用、或对某一种疾病起作用。所有的药物都是有副作用的。开发一种新药十分困难。当药物分子进入病人的体内后，它们必须能在成千上万种不同的分子中找到某些特定的分子并与它们相结合，从而使病人摆脱这些分子对他们造成的病痛与折磨。

药物必须是一种对我们的身体有益且尽可能无害的分子，同时亦必须能给制药公司带来财富。当下，我们的社会普遍认为公共卫生是一个赚钱的事业，而不是社会的职责。当药物开发人员筛选出一些候选分子时，他们首先将其用于小鼠、大鼠、仓鼠、狗等动物进行试验。只有在动物身上试验成功的药物才会被进一步用到人的身体，进入"临床试验"。这是一个十分谨慎、严格管制、缓慢且昂贵的工程。当然，大多数临床试验是不成功的，其主要的原因在于药物在人体内的作用方式有可能与其在老鼠体内大相径庭，这使得很多制药公司对新药研发的投入不断萎缩。虽然哺乳动物都有着类似的代谢机制，但不同的物种却有着本质的差异。那些能治愈老鼠疾病但却不能为我们人类所用的药物只能像垃圾一样被丢弃。

如果临床试验的人力物力投入太大、周期太过缓慢、同时又受到道德和法律的约束，那我们应该怎样去试验一些分子是否对人体有效和安全呢？一种办法就是利用从人体组织或血液中分离出来的细胞作为被测试对象。虽然人体细胞并不能代表完整的身体，而且细胞与有机体之间存在着天壤之别，但无论怎么讲，人体细胞内确实包含全部的人类基因组和人体生化过程的本质特征。

观察细胞有多种途径。这张图片是用荧光显微镜拍摄的，不同颜色的染料向我们展示了细胞的内部结构。当细胞生物学家看到这个结构时，他（她）能大概估计出这个细胞的喜怒哀乐，甚至能推断出被测细胞对外来的分子所作出的反应。这个外来分子可以是药物研发者们辛辛苦苦筛选出来并希望它能治疗人们疾病的新药。

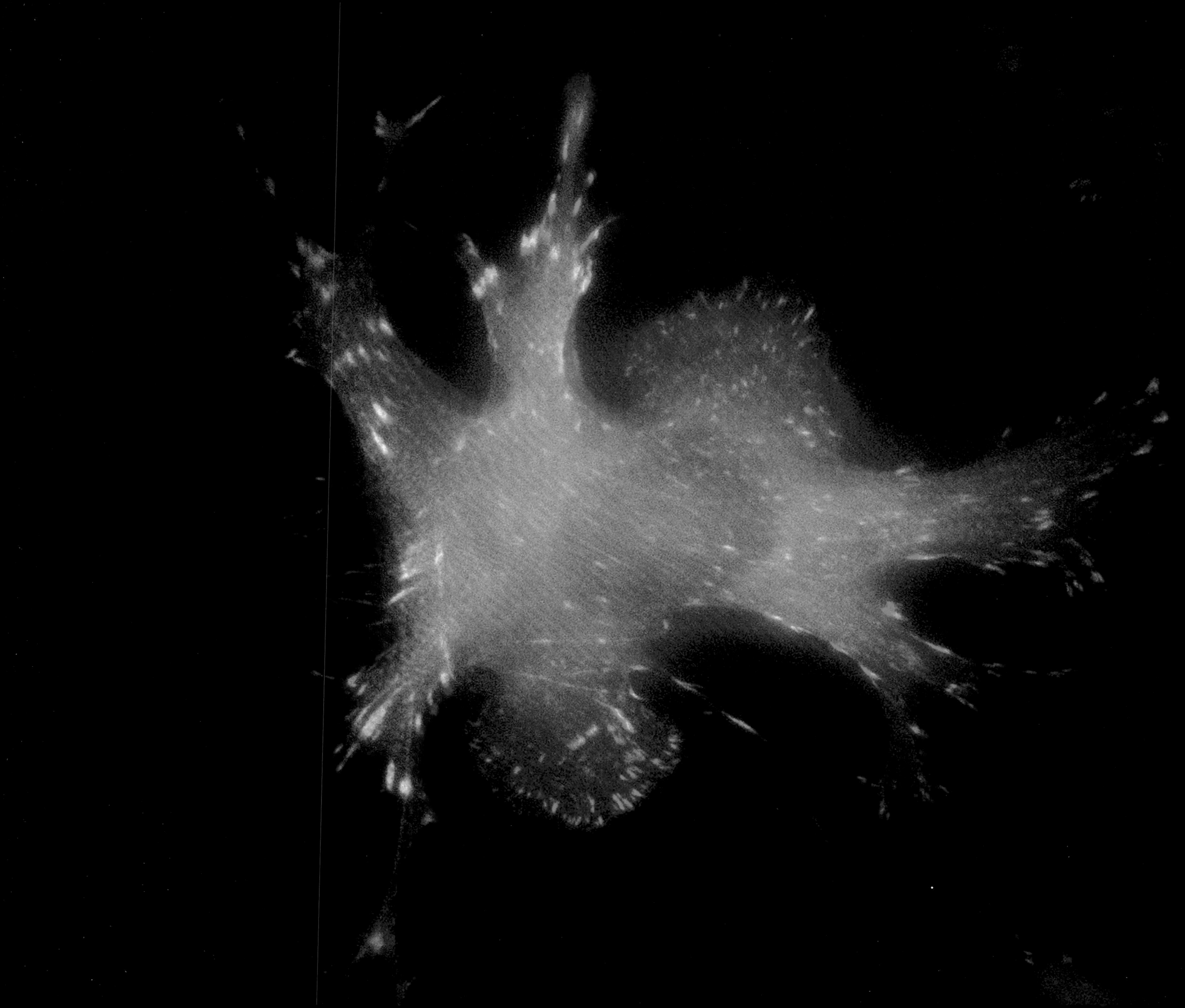

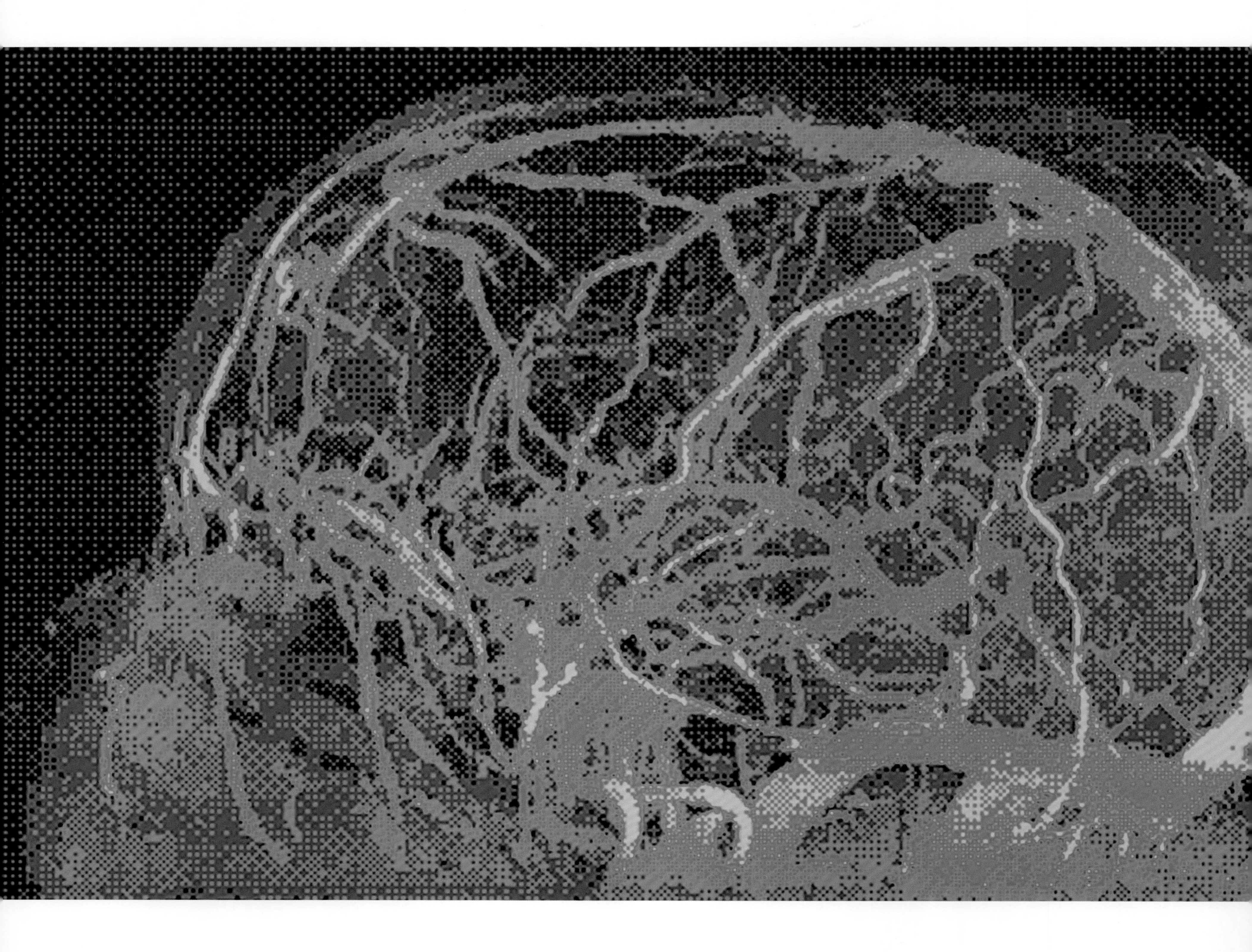

给发烧的大脑降温

Cooling the Fevered Brain

47

大脑和计算机都是用来处理信息的，但除此之外，两者迥然不同。计算机达人通常蔑称大脑为“湿件”，它包含组织、血液，以及各种各样的生物组织和介质；而计算机则是一种“硬件”，它由硅、陶瓷和金属组成。大脑依靠离子浓度、蛋白质结构以及细胞之间的相互作用来存储信息，至于它如何工作的，我们还知之甚少。相比之下，计算机却以一清二楚的原理将信息存储在电子和磁场中。大脑能聪明地识别图案；而计算机却能快速地进行数字运算。大脑让我们在愉悦地享受烛光晚餐时迸出爱情的火花，而计算机却能让我们用信用卡来为这顿晚餐付款。

大脑是靠液体来提供营养并保持恒温的，这种液体就是血液。一个思考中的大脑会持续升温，即使仅仅升高几度，也会让人变得木讷，如果再升高几度，那大脑就会变成“炒熟的鸡蛋”。大脑中盘根错节的网状血管以及在其中流动的血液为大脑提供足够的糖分和氧气以维持其功能，并保证大脑不至于过热，以及除去新陈代谢反应的产物。

核磁共振成像（MRI）能用来监测大脑中网络状的血管，如果在血液里注入一些氧化铁纳米颗粒，图像会变得更加清晰。这些网络状的血管对那些想知道大脑是如何工作的人特别有用。当大脑出现故障时，核磁共振成像也能够为大脑的修复工作提供帮助。

通过核磁共振成像技术，可以观察到脑内是否有不正常的血管，进而用来检测脑部肿瘤。在不久的将来，这项技术或许还可以用来测谎。值得庆幸的是，这项技术目前还不能用来揭穿一个人谈情说爱时的甜言蜜语。

BRAINS PROCESS INFORMATION; so do computers. Beyond that similarity, the two are mostly different. Brains are what computer geeks disparagingly call "wetware": tissue, blood, and generally messy biological stuff. Computers are "hardware": silicon, ceramics, and metal. Brains store information in concentrations of ions and, in ways we only foggily guess, in proteins and connections between cells; computers store information in electrons and magnetic fields using rules we understand completely. Brains cleverly recognize patterns; computers rapidly add numbers. Brains make it possible to fall in love over dinner; computers let our credit cards pay for dinner.

Brains are liquid-fed and liquid-cooled; the liquid is, of course, blood. A thinking brain heats up. If its temperature rises only a few degrees, it begins to stutter; a few degrees more and it turns to scrambled egg. A network of blood vessels and flowing blood supplies it with the sugar and oxygen it needs to function, protects it from overheating, and removes the waste products of its metabolism.

Magnetic resonance imaging (MRI) can trace the network of blood vessels in the brain; adding nanoparticles of iron oxide to the blood sharpens the images. This network of pipes is riveting to those curious about how the brain thinks, and sometimes helpful to those who try to repair it when it breaks down.

One way to search for an aneurysm or a tumor is to use MRI to look for characteristic irregularities in the vasculature of the brain. In the future, it may also betray whether a brain (or the person it constructs) is lying or telling the truth. Fortunately, watching a brain fabricate a romantic compliment over dinner is still impossible.

A CHEETAH IN THE UNDERBRUSH?

We startle easily. A mildly surprising beginning–a sound out of place, a shadow where none should be, too many tics from a Geiger counter–can have a starkly lethal end.

"Unexpected" can be good or bad. We humans assume "bad," and with excellent reason. "Unexpected and good" we can adapt to at our leisure in the future; "unexpected and bad" can leave us with*out* a future.

Darwin would say it's smart to be conservative when facing something new.

灌木丛中的黑影？

我们很容易被惊吓。不合时宜的声音、荒郊野外的黑影、甚至盖革计数器（测量放射线用）上过高的读数……原因很简单，一个小小的意外往往会导致十分严重的后果。

"意料之外"的事或许是惊喜，也可能是噩耗。"惊喜"可以在未来闲暇时慢慢地适应，而"噩耗"却有可能让我们万劫不复。因此，人类完全有必要在"意外"来临时做好最坏的打算。

正如达尔文所说，对新事物采取保守的态度是最明智不过的选择。

Phantoms

幽　灵

48

Early on, our species realized that we held a weak hand in the card game of evolution. If we met a cheetah, the cheetah won. If we had a rock or a sharp stick, the cheetah still won. Darwin would have bet on cheetahs.

For once, Darwin might have bet wrong. We learned, eventually, that by forming groups with many sticks and many rocks, and by exercising our anatomical oddities—a large brain, a larynx that formed the complex sounds of language, an opposed thumb—the odds switched to favor us. Still, the unpleasant experiences of the childhood of our species have tended to stick. Faced with something unfamiliar, we are simultaneously curious and afraid. To remind us to be careful, we construct things that go bump in the night—phantoms, trolls, chain-saw movies.

This field of shimmering light, with unrecognizable shapes inside, could be anything: best to assume it's dangerous! (It is, in fact, a fountain playing over rocks; but no matter.) We treat new technologies in the same way. At the beginning, they are all dangerous: fire, the book, steam engines, genetically engineered bacteria, nanotechnology. But with enough familiarity, we ignore even the truly dangerous ones—nuclear weapons, ubiquitous surveillance, smoking, drinking and driving, the sports channel on TV.

And even after learning how to deal with cheetahs, we might neglect to notice that the stick on the road is a black mamba, or that a 20 percent annualized, risk-free, after-tax return on investment has a funny smell.

Perhaps Darwin would have bet right after all?

在进化的早期，人类就意识到在生存的游戏中我们始终处于劣势。当我们碰到猎豹时，不管有没有石头或尖锐的棍子作为武器，我们都会输。如果把这场较量看作是一场赌博的话，达尔文肯定会在猎豹身上下注。

倘若如此，达尔文就下错赌注了。因为人类懂得学习，我们不会傻到去跟猎豹单挑，相反，我们会号召所有手持棍棒和石头的人组成队伍共同作战。我们还可以发挥聪明脑袋的优势，通过各种语言相互交流，借用各种手势相互配合。渐渐地，人类成为地球上的主宰，包括猎豹在内的猛兽都成为手下败将。但是，童年的不愉快往往会一直折磨我们到老。当我们碰到不太熟悉的东西时，心里总是充满好奇与害怕。我们甚至还编造出各种各样的鬼怪故事，以提醒自己时刻注意。

作为一个新兴的领域，纳米技术究竟能带给人类什么，现在还不得而知。最好的办法就是假设它是有危险的，我们需要多加小心。我们也以同样的心态来对待其他新技术。刚开始的时候，它们都是十分危险的，比方说火、书籍、蒸汽机、细菌的基因工程以及纳米技术。但当它们被逐渐认识和了解，甚至像枕边人那样熟悉后，一切就不再可怕了。我们总是害怕新事物，而忽略那些真正危险但已习以为常的东西或行为，例如核武器、醉驾、抽烟、电视上的体育频道以及无处不在的监视。

当我们知道怎么对付猎豹时，可能会对地上的一根"棍子"疏于防范，而实际上那是一条剧毒的黑曼巴蛇。同时，对于一项无风险且收益高达20%年利率的投资项目所散发出来的诱惑，我们往往也会难以抗拒。

由此看来，达尔文最终还是下对了赌注？

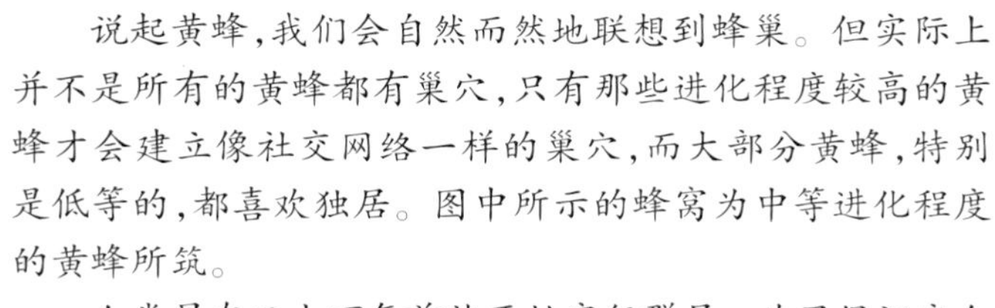

隐私与蜂窝

Privacy and the Nest

49

说起黄蜂，我们会自然而然地联想到蜂巢。但实际上并不是所有的黄蜂都有巢穴，只有那些进化程度较高的黄蜂才会建立像社交网络一样的巢穴，而大部分黄蜂，特别是低等的，都喜欢独居。图中所示的蜂窝为中等进化程度的黄蜂所筑。

人类早在二十万年前就开始实行群居。为了保证安全与相娱，大家以团体的形式生活在一起。人类的历史就是一部族群的壮大史，经历了从家庭、部落、村庄到城市、国家，甚至联盟的变革。随着技术的进步，我们还建立了更大的地球村。其工作模式与原始社会相比并无太大差异，只是规模日益壮大。

人是爱唠叨的动物，喜欢交谈。起初，通讯基本靠吼。后来，我们发现串串门子吃吃饭，顺便谈论一下家长里短也是不错的生活方式。随着时代的发展，报纸、电视、电话、高速公路、飞机等的出现为我们交流提供了新的途径。现如今我们又发明了因特网，这是一种可以提供数也数不清的社交方式的新媒介。黄蜂用纸一样的材料搭建了它们的巢穴，而我们则用光纤、电脑和比特（信息的单位）建成了我们的“安乐窝”。

网络是把双刃剑，它拥有的海量信息既能丰富我们的生活，也会威胁到我们的隐私。如果人人都能通过网络找到关于某人的任何信息，那么这把剑的弊端就显露无遗。这意味着隐私和个性的终结吗？我们当然希望这个不会发生。在这种情况下，“六度空间理论”可以工作得更为有效。试想一下，当我们把计算机、人以及网络结合在一起时，就能更有效、更快捷地帮助我们找到任何问题的答案。这样一来，我们就会跟周围的人自动保持距离，因为我们所知道的和所了解的东西都一模一样。

与之交流还是保留隐私，其实你并无选择！

When we think "WASP," we think "nest." But in fact, most kinds of wasps—especially the more primitive ones—are solitary; only the more advanced have invented the complex social world of the nest. This nest was home for a species of intermediate complexity.

We humans learned in our dawn—perhaps 200,000 years ago—that groups offered safety and company. Our history has been one of ever larger groups—family, tribe, village, city, nation, coalition. Technology is just now allowing us to build a yet larger group: the global village. Back to the beginning, but much, much bigger.

Humans chatter: we love to talk. At first, we shrieked and hooted across the savannah. Later, we walked to the next village to share supper and news. Newspapers, highways, telephones, airplanes, TV—each provided a new way for the village to gossip. Now we have invented the Internet, a new kind of nest for a more complex society. Wasps build their nests of paper; we build ours of optical fiber, computers, and bits.

Gossip is terrific fun, and good for business. It also threatens privacy. If everyone can find out anything about anyone just by asking the Web a few questions, we've gained connection but lost separation. Is the end of privacy also the end of individuality? We hope not, but "six degrees of separation" really works: the combination of computers, humans, and the Internet is startlingly close to being able to answer any question we can ask. And if we all know the same answers to the same questions ...?

Do we like this thinning boundary between connection and privacy? Do we have a choice?

Soot and Health

灰尘与健康

50

SHOULD WE WORRY ABOUT nanoscale things? Is there anything about them that should cause alarm?

The lungs are the swinging door into the body. All small things are ... well ... small, and tend to float in air, where we inhale them. If deposited in the lungs, these particles might, perhaps, be eaten by cells, and the smallest–those approaching the size of molecules–might even pass between cells and into the blood.

None of which is new. Our species has always had lungs, and we have always lived in an unseen fog of small particles suspended in the air we breathe–in the smoke from wood fires, in the brisk air of the seaside, in the soot from the back of a diesel bus, in dust everywhere. Air may look transparent, but it is usually laden with particles too small to see.

What about a special toxicity of small particles? Unexpectedly, it is probably not the tiniest particles that are the greatest concern. Real nanoparticles tend not to deposit in the lung: we breathe them in, and we breathe them out again. Somewhat larger particles (a few microns across) stick more efficiently. In any event, we don't yet know the answer to this question. Certainly some particles (chalk, silica, asbestos, coal, cigarette smoke, urban pollution) can injure lungs. But whether there is something special about the smallest particles we breathe remains to be discovered.

我们需要担心纳米尺寸的物体吗？它们又有哪些特性值得我们警惕呢？

肺是进入身体里的一扇门。当我们呼吸的时候，所有小到能够漂浮在空气中的颗粒都有可能通过这扇门侵入我们的身体。一旦这些颗粒在肺里沉积下来，它们将有可能进入细胞，小到接近分子尺寸的颗粒甚至还可以穿过细胞间隙进入血液。

所有这些都不是什么新鲜东西。我们人类自古以来就有肺，并一直生活在布满了悬浮颗粒的空气中。这些小颗粒可能来自森林大火的浓烟、风化的岩石、柴油机车的尾气以及无处不在的灰尘。空气看起来虽然是透明的，但里面总是有很多肉眼看不见的小颗粒。

小颗粒有没有特定的毒性呢？它们是否会威胁到我们的生命和安全呢？总的来说，到目前为止我们仍然不知道这个问题的确切答案。但出人意料的是，真正需要关注的并非是纳米级的小颗粒，我们能将它们吸入，也能将它们呼出，因此它们一般不会在我们的肺里沉积。反而是那些微米级的颗粒，例如粉笔灰、二氧化硅、石棉、煤尘、烟灰、城市灰尘等才会吸附在肺叶上排不出来，从而伤害我们的肺。至于那些纳米级的颗粒物是否会在被吸进之后对人体产生一些特殊的作用，还有待进一步研究。

机 器 人

Robots

51

什么将成为明日之星？信息科技已将世界各国变成一个地球村，生物科技正开始发挥作用，纳米科技也已从襁褓中迈步而出，那这些之后呢？

有一种可能是机器人。下一页所展示的音乐盒就是一个简单的机器人，它利用橡皮带上装有发条的马达来推动齿轮转动，从而代替弹奏大键琴的纤纤细指。现今的机器人更为复杂，但仍不够灵活（如升降机和电焊机），而且大多数都是固定在某个地方。好在我们有信息技术，它正在赋予机器人漫步行走的本领。例如，脱粒机和全球定位系统的结合已令收割机可以自己在田间穿梭；无人驾驶飞机则可以在千里之外进行遥控；机器苍蝇（左图）则是一款新型品种，它是一种能执行复杂的机翼运动并进行盘旋的微型机器人。

一个可移动的机器人必须具备判断能力。如果一个机器人在碎石上滑倒，或遭遇到湍流，它必须立刻作出反应，而不是跟遥控人员闲聊之后再制订对策。聪明的设计人员已开始让机器人拥有一套跟动物一样的条件反射系统，也就是一个能快速测量、决断并行动的局部电路。

机器人将如何进化？我们还不得而知。但如果跟信息技术相结合，这个答案将会是“非常快速”及“无法预料”。我们需要机器人接过那些让人厌烦的甚至危险的工作。它们最终会发展成什么样呢？是自己能调节的剪草机？像螳螂一样具有快速反应能力但没有个性的莽夫？还是拥有新型智慧的非生命系统？谁知道呢？

人类的进化遵循着类似的途径。不管是人类还是那些仅具有简单生命的有机体，靠的都是“适者生存”的准则。如果没有那么多代的繁衍、进化和智慧的进步，历史上就不会有简·奥斯丁和孙武这样杰出的人物。进化是如何使我们越来越聪明的呢？我们有很多的猜测，但还是没有结论。

我们靠工作来维系生计，而机器人比我们更廉价、更娴熟。如果它们来跟我们竞争这些工作，那我们该如何应对？我们迟早要面对这种考验，虽然不是现在，但一定会是在不远的将来。

WHAT IS THE *next big thing*? Information technology has collapsed a world of nations into one global village. Biotechnology has begun to make a difference. Nanotechnology is graduating from diapers. And after that?

One possibility is robots. A music box (next page) is a primitive robot: prongs on a rubber belt moved by a wind-up motor replaced the delicate fingers that played the harpsichord. Today's robots are more sophisticated but still clumsy: lifters and welders. Most are fixed in place. But information technology is now giving robots their freedom to wander. A marriage of threshing machine and global positioning system has produced harvesters that guide themselves across the fields. Airborne drones obey controllers thousands of miles away. The robotic fly (opposite) is an early representative of a new species: a microrobot able to execute the complex wing movements necessary to hover.

A mobile robot must be able to make its own decisions. If, as a robot, you slip on gravel, or encounter a wind shear, you must react instantly; you cannot have a leisurely chat with a distant advisor. Clever designers build reflexes, like those in animals, into their robotic offspring—local circuits that quickly measure, decide, and act.

How will robots evolve? We don't know, but if information technology provides a guide, the answer is "*very* rapidly" and "in ways we cannot anticipate." We *need* robots—to relieve us of unpleasant, boring, or dangerous jobs. But what will they become? Adaptive lawnmowers? The mechanical equivalent of the praying mantis—with lightning reflexes but no personality? Nonliving systems that house a new kind of intelligence? Who knows?

We do know that we followed a similar path in our own evolution. We—or better, the first primitive organisms—survived as collections of reflexes. Over not so many generations, reflexes, intelligence, and Darwinian selection produced Jane Austen and Sun Tzu. How, or if, reflexes aggregate and become intelligence is a subject about which we have many opinions, but understand almost nothing.

Regardless, there will be problems—not now, but sometime. We humans *are* our work, and robots will compete with us for that work. If they are cheaper than we, or more skilled, how will we spend our days?

Fog

雾

52

"FOG" IS "CLOUD" UP CLOSE—drifting translucent shadows that blur the ordinary into childhood fairy tales. Fog is a haze of finely divided particles of water or ice suspended around us; when light enters, it refracts at the surfaces of these particles. The blinding white of a cumulus cloud at summer midday is the reflected light that does *not* make it into the cloud's interior. When we are outside, it's "cloud"; inside, it's "fog."

Fog from water is familiar, but any transparent liquid will do. Los Angeles smog is made up of hydrocarbon droplets—residues from gasoline burned in cars and trucks. Appalachian blue haze consists of similar hydrocarbon droplets exhaled by trees. When it comes to fog, "loathe" or "love" is a delicate question of smell and context.

A particularly interesting kind of fog might form if we injected sulfuric acid into the upper reaches of the atmosphere. It would form a thin veil and cast a faint, cooling shadow on the planet below. Technically, veiling a planet in man-made fog would be laborious, but conceptually it is straightforward: enough rockets will do it. And we know the idea works: an Indonesian volcano—Tambora—erupted in 1815 (the largest eruption in the last few thousand years) and coughed enough ash into the atmosphere to shadow and cool the globe. This cooling produced New England's "year without a summer" in 1816.

And why would one want to cool the earth? If the planet becomes too hot—if it seems that global warming is getting out of hand—perhaps we might *need* to cool it? Perhaps we might, and perhaps we also might not. A high-altitude fog might cool the globe, but it might cool it too well. We should be careful when experimenting with the only planet we have.

所谓的"雾"是一种我们能近距离接触到的"云",它是漂浮在我们周围的半透明阴影,让人仿佛置身于仙境一般。雾是一种由水或冰的细小颗粒所组成的悬浮体,在光线照射下,表面的颗粒物会将光反射回去。我们在夏天所见到的、白得耀眼的云朵正是由于阳光没能穿透云层而被表面反射回去所造成的。因此,当我们从外面看它时,它就是"云";而当我们置身其中时,它就成为"雾"。

我们所熟悉的雾大部分由水滴形成,当然也可能由其他透明液体构成。例如,洛杉矶的烟雾就是由汽车尾气中的碳氢化合物液滴所形成的;阿巴拉契亚山脉的蓝色薄雾则是由树木释放出的一种类似碳氢化合物的液滴所组成。对于雾的厌恶和喜爱,我们的嗅觉和视觉往往会给出十分微妙的反应。

如果我们将硫酸发射到大气层的上方,就可以形成一种特别有趣的雾。它会像一层薄纱一般笼罩着地球,并投射出微弱的、冰冷的阴影。从概念上来讲,这件事并没有什么难处,只要有足够的火箭便可;可从技术上讲,用一层人造的薄雾将地球笼罩虽可行却还是很费力气的,但这对于大自然来说却是小菜一碟。例如,1815 年印度尼西亚的坦博拉火山爆发(过去几千年中最大的一次火山爆发)并释放了大量的火山灰到大气层中,使地球变得阴暗和寒冷。这一事件导致的急剧降温使新英格兰在 1816 年都不曾有夏天。

我们为什么要给地球降温呢?试想,如果全球温室效应恶化到了无法被控制的地步,我们还能找到一片清凉的绿洲么?或许我们可以选择利用高空大气层中的雾来有效地降低地球的温度,但是千万别将温度降得过低!在对人类赖以生存的唯一星球进行实验时,我们必须格外小心。

疾病与健康

In Sickness and in Health

53

几千年以来，瘟疫就像一个手持大剪刀的园丁，如修枝剪叶般对地球上肆意增长的人口进行削减。究其根源则是那些看起来又小、又笨、又没爱心的细菌或病毒。

如今，细菌甚至比手表或者发条玩具更容易进行重组。在适当的指导下，即使是高中生也能通过基因工程技术在细菌内植入新的“部件”。这种能力在医药领域弥足珍贵。当生物学家们发现某种蛋白质对人类有用时，他们会通过构建特定的细菌来大量生产这种蛋白质，希望以此来增进人类的健康。

但是，如果带着反社会的意图去重组细菌将会产生什么后果呢？比方说，从一种病原体如瘟疫、霍乱出发去制造毒性更强的细菌与病毒，并且使它们对抗生素有更强的抗药性，这又会带来什么样的灾难呢？

理论上，专业的基因学家可以进行这样的尝试，对他们来说，困难的不是用 DNA 来写指令，而是写何种指令。虽然人类自从诞生以来就与疾病为伴，但直到现在我们仍然不知道这些古老的宿敌到底是什么，而对于怎样去发展它们更是知之甚少。与细菌产生的有毒蛋白相比，疾病更为复杂难懂，因为它是病原体和宿主相互作用的结果。它们的毒素，是我们的抗菌素；它们的附着因子，则是我们细胞表面的分子。

正是由于目前对疾病的规律知之甚少，对一个新病原体的危害也很难估计，使得我们只能在一定程度上保护自己。我们必须认真监视人们对病原体基因的改组（也就是纳米工程在生物中的应用）。那些懂得这方面技术的人还是可能给人类构成威胁的。

可悲的是，我们根本不需要用任何新的病原体来对人类造成伤害。炭疽、天花、兔热病、鼠疫等等这些熟面孔即使没有经过基因改良就已经给人们带来了无尽的痛苦。

FOR MILLENNIA, PLAGUES HAVE BEEN the pruning shears that trimmed unruly growth in population. It's annoying that the gardener is small, mindless, and uncaring—a bacterium or virus.

Bacteria are now easier to re-engineer than watches or windup toys. With guidance, high school students can fit new parts into living cells through genetic engineering. This new competence is invaluable in medicine: when biologists discover proteins that do something helpful in us, they can often construct bacteria to make these proteins in quantity, with the intention (and occasionally the result) of improving our health.

But what about antisocial intent? What about trying to *cause* disease rather than cure it? What about starting with a pathogen—plague or cholera—and making it more resistant to antibiotics, perhaps, or more virulent?

In principle, any skilled geneticist can try. The difficulty is not writing instructions in the DNA, it's knowing what instructions to write. Although we have lived with these ancient enemies for as long as we have been a species, we still barely know what "disease" is, much less how to engineer new features into it. Disease is more complicated than just a few toxic proteins produced by a bacterium; it's a tango that hosts and pathogens dance together. *Their* toxins, *our* antibodies; *their* attachment factors; *our* cell-surface molecules.

So we are protected—to a degree—by our current ignorance of the clockwork of disease. Also by uncertainty about what a really new pathogen might do. Still, genetic modification of pathogens—nanoengineering applied in biology—is something to watch very, very carefully. People with a grudge can still be competent.

But, sadly, nothing new is really needed to cause harm. The old standbys—anthrax, smallpox, tularemia, plague—are more than good enough, given the right opportunity, and without genetic "improvement," to cause endless misery.

WHALE OR HERRING?

Is nano a change in the world-as-we-know-it, or a passing trifle?

Our forebears, when they set out to make their fortunes on the briny deep, had a choice: try to catch a very occasional whale, or seine for herring. Which is nanotechnology? "Whale," and change-the-world, or "herring," and reliable-for-dinner?

We don' t yet know.

鲸鱼还是鲱鱼?

"纳米"技术将彻底改变我们的世界,还仅仅是过眼烟云?

我们的祖先在去深海捕鱼时可能也会有同样的纠结,是去捕获能改变命运但偶然一现的鲸鱼还是休闲地撒下渔网围捕鲱鱼? 对于我们来说,哪一种更像是纳米技术?是意味着改变命运的"鲸鱼",还是能保障晚餐的"鲱鱼"?

目前我们还不知道。

The Internet

因特网

54

PHYSICALLY, THE WORLD-WIDE WEB—the Internet—is a network of information processors, connected by antennas and optical fibers. It is perhaps the most ambitious project we have started as a species. It is the largest thing we have ever built, and we have assembled it from transistors—the smallest things we know how to make (four billion can fit on a postage stamp). It is a chrysalis we are forming around the planet. Or perhaps it is forming itself. We suspect that something new will emerge from it, but we don't yet know what.

Functionally, the Internet allows information processors to exchange tsunamis of bits, most incarnated as pulses of light and electrical current and bursts of microwave transmissions. The hubbub! A part of the world—the richest part—already lives in an atmosphere that is a froth of air and bits. The poorer parts are catching up: first water, food, and electricity, then cellphones and the Internet. The Internet is a table where we sit to gossip; a *suq* where we buy and sell; a shadowy corner for planning mischief; a library holding the entire world's information; a friend, a game, a matchmaker, a psychiatrist, an erotic dream, a babysitter, a teacher, a spy. It is what we need it to be. The best and worst and most ordinary of us, reflected—and perhaps distorted—in a silvery fog of bits.

We humans love to talk to one another. That's why we invented speech, writing, books, telephones, TV, travel, and now the Internet. These images tell the tale for the United States. The upper one maps an old technology: the flow of passengers on airplanes. The lower one maps the new: the flows of bits.

The various parts of our bodies also have to talk among themselves—that's how the brain, and spinal cord, and nerves evolved. How do our internal biological and external electromagnetic systems of communication compare? Evolution invented us, and then we became the multiple parents of the Internet. But postpartum, the Internet is evolving itself. It has *many* "brains"—ours and our computers'. In creating the world-wide web, we have not replicated ourselves on a planetary scale: we, and it, are entirely different. And so far, the web has not developed a "brain"—a central controller.

So far.

在物理学上，万维网(即因特网)是一个由信息处理器、天线和光纤连接而成的网络。因特网也许是人类有史以来最宏伟的工程，但它却是由我们所能制造出的最小东西——晶体管一个一个组装而成的。要知道，40亿个晶体管有序地组装起来才相当于一张普通邮票的大小。因特网就像一只将全世界囊括在内的蝶蛹，我们知道有些东西即将破茧而出，但却不知道那个飞出来的东西会是什么。

因特网的功能是让海量的信息化身为大量的光电脉冲和爆炸式的微波信号得以快速传输。世界因此变得沸反盈天！富人那端的生活已经进入了信息化时代，连空气中都充满了数据。而穷人这端则在奋起直追，起初只有水和食物，接着是电器，然后才有手机和网络。因特网的出现，给整个社会带来了巨大变革。它是一张可以让我们坐下来闲聊的桌子，是能买卖物品的露天广场，是策划恶作剧的阴暗角落，是拥有海量信息的图书馆；它是我们身边的朋友、媒人、医生、保姆、老师、密探或是一场游戏、一夜春梦……我们希望它成为什么，它就是什么。无论是最好的、最差的、还是最普通的，我们都在因特网这层银色薄雾的笼罩下生活着。

人类喜欢交谈，所以我们发明了演讲、写作、书籍、电话、电视、旅游以及现在的因特网。右图所展示的是两幅特殊的美国地图：上面一幅由往来各地的飞机航线所组成，体现的是传统技术；下面一幅则由城市间网络信息的传输绘制而成，它代表了现代技术。

我们身体的各个部位也会相互“交谈”，大脑、中枢神经和周围神经正是通过这些“交谈”而得以进化。那么，我们身体内部生物系统间的交流方式与外部电磁系统间的交流方式有可比性吗?进化创造了人类，现在我们又创造了因特网。但是自打出生起，因特网就在进行着自我进化，我们及计算机都在充当着它的“大脑”。然而，因特网并不是我们的复制品。迄今为止它还没有形成自己的“大脑”。

不过，这仅仅是目前的状况而已，将来怎样还得拭目以待。

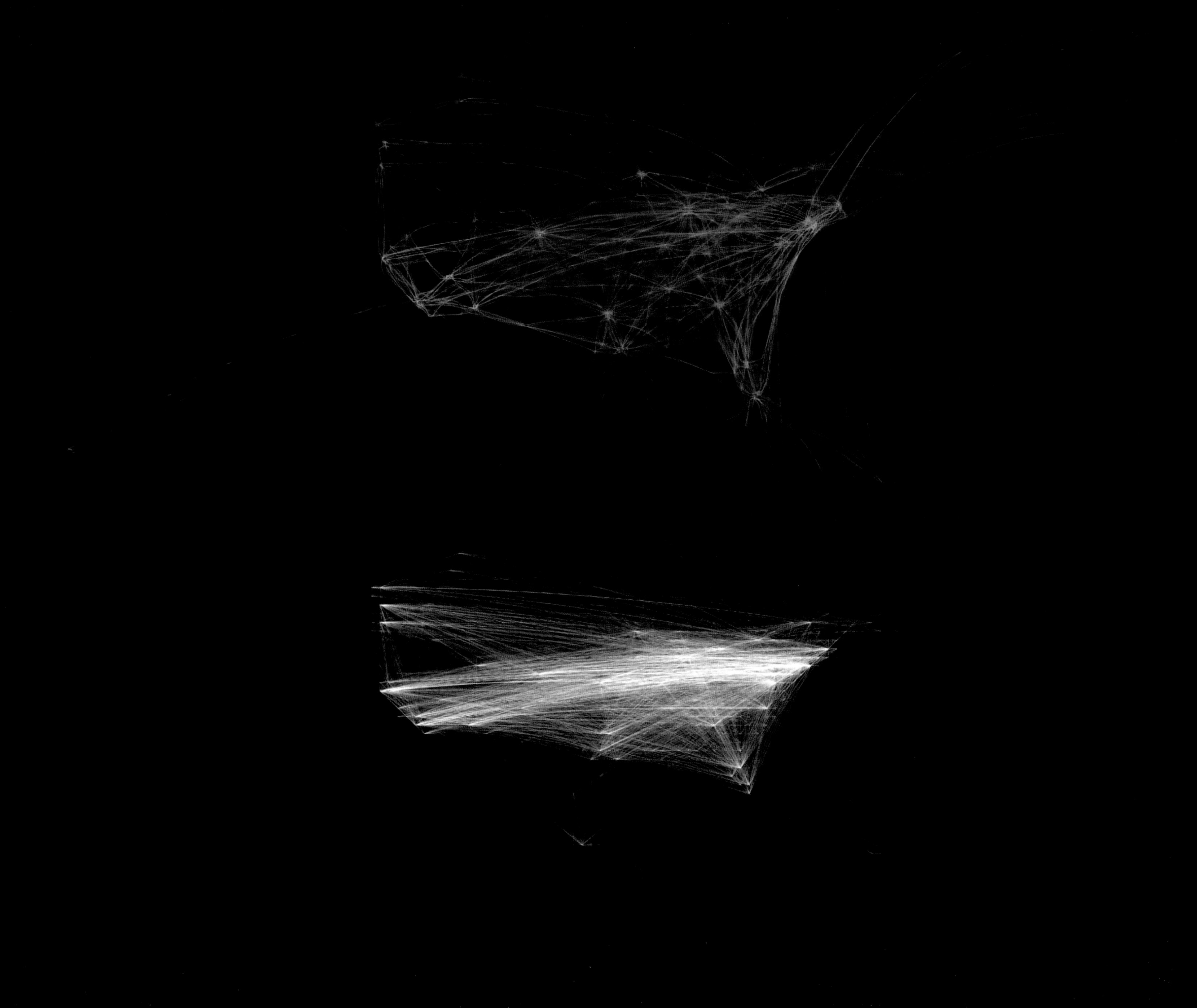

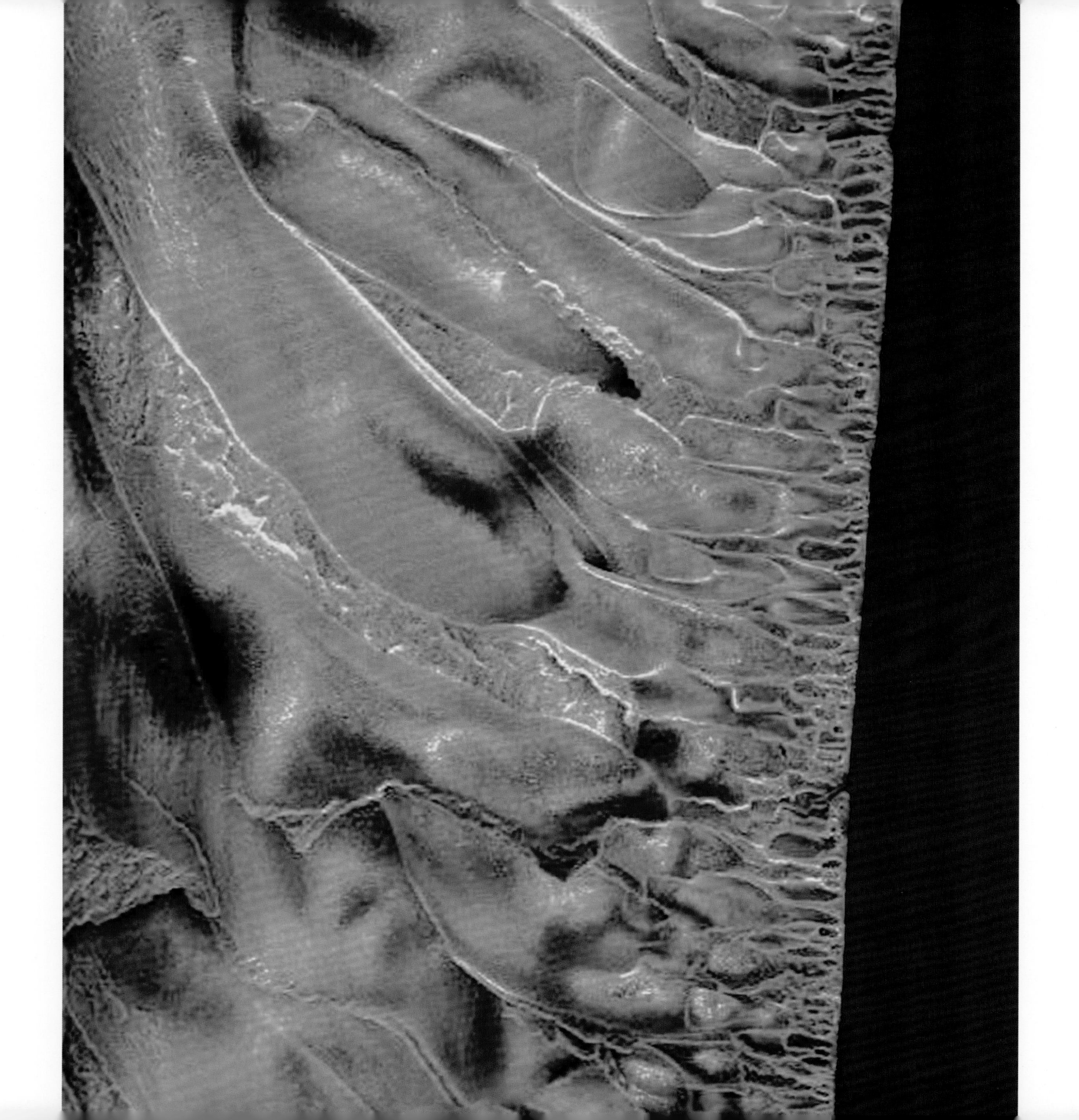

反渗透膜

Reverse Osmosis Membrane

55

我们人类就像漂浮在水上的一艘船，没水寸步难行。特别是纯净水，没有它的话我们就会像那些贫穷不幸的人们那样遭受肠道微生物所带来的疾病，如霍乱、痢疾、血吸虫病、贾第虫病、隐孢子虫病等等。这些古老的肠道微生物已经和我们一起生活了上千年，大多数实际上就寄生在我们的体内。只有拥有了纯净的水，我们才能远离疾病，建立一个繁荣、稳定的社会。

如果有一条淡水河或一片湖泊，要想得到纯净的水相对就比较容易。通过过滤去除水中的沉淀以及异味和有毒之物，再加入少量的氯气或者臭氧以消灭病原体，这样的水就饮用无妨了。但是，如果我们只有咸水，就必须先让它通过一张反渗透膜，将其中的盐分离出去。

这些膜就像是玄关处的纱门，只不过它的孔径是纳米级的。在夏日里，纱门可以挡住苍蝇和蚊子，但小蠓虫仍然能畅通无阻。水分子和氯化钠(盐)的离子虽然在尺寸上相近，但却有着截然不同的电荷性质，前者不带电，后者却带电。反渗透膜只能让水分子通过而带电的离子则不能。

这张图片展示了一张反渗透膜的横截面，其复杂的内部是由一系列的小孔构成的。孔径大小的选择是以膜的强度和韧度达到最佳为准则。为了让水分子能顺利地通过这张带孔的膜，同时阻挡住盐分，通常需要施加相当于标准大气压1000倍的压力来完成。然而，产生这样大的压力就必须要消耗大量的能源与财力，而且太大的压力也容易使本来就脆弱的膜破裂。正所谓，天下没有免费的午餐。

THE MIDDLE CLASS IS A SHIP that floats on clean water; water is the fluid foundation of societies. Without clean water, we share the intestinal organisms that are our fellow travelers in poverty: cholera, dysentery, schistosomiasis, giardia, cryptosporidium—all the old, old companions who have been very close to us (in fact, inside us) for millennia. Only with pure water can we build a prosperous, stable society.

If a fresh-water river or lake is available, it is relatively easy to get pure water. Filtration removes sludge, taste, and toxins. A touch of chlorine or ozone destroys the pathogens. Then drink! If the only available water is salty, a separate step is necessary to take out the salt: passage through a reverse osmosis membrane.

These membranes are functionally the nanoscale analogs of porch screens on a summer evening. Screens block flies and mosquitoes but not biting midges. Molecules of water and ions of sodium chloride (salt) are not so different in size but are very different in electrical charge: water has no charge; sodium and chloride ions are electrically charged. A reverse osmosis membrane bars the ions but allows the water molecules to slither through.

This image is the cross-section of an reverse osmosis membrane; its complex internal structure is a series of pores of sizes chosen to maximize its strength and durability, while simultaneously supporting a thin surface film that accomplishes the ultrafiltration. To force water through this membrane, and to separate pure from salty water, requires pressure—almost a thousand times the pressure of the atmosphere in which we live. Generating pressure requires energy and costs money, and can rupture a fragile membrane. There's no free lunch.

Nuclear Reactions

核 反 应

56

THE AMOUNT OF ENERGY that is available from chemistry—by combustion, from fuel cells and batteries—is limited by the characteristics of atoms and molecules. You can only get so much out of chemical reactions. We want *more*—more energy and cheaper energy—ideally, something for nothing. Is there nothing better?

Indeed there is: *nuclear* reactions. Rather than making new molecules by shuffling nuclei and electrons using chemistry, we can make certain nuclei shuffle protons and neutrons and form new *atoms*. The energies available from nuclear reactions are, perhaps, a million times greater than those available from chemical reactions. Unfortunately, the nuclear fuel must contain isotopes that fragment in a controllable way; uranium, plutonium, and thorium are the best, but all are rare and expensive.

And there are other problems. One is that uranium and plutonium make both power plants and nuclear weapons. A second is that most people don't want to live next to a nuclear power plant, very much don't want a nuclear waste site in the neighborhood, and most emphatically don't want to disappear in the transient sun of a nuclear explosion.

Nuclear power is, and will be, an essential contributor to the commercial production of energy. Beyond nuclear fission lies nuclear fusion—processes whose practical potential for producing power has been debated for decades—and beyond those, a range of even more energetic subatomic processes. This image shows—in a forest of other traces—evidence of the collision of pions (members of the zoo of exotic particles that fascinates high-energy physicists) with hydrogen atoms, written as strings of bubbles of hydrogen gas, in liquid hydrogen, at a temperature a few degrees above absolute zero.

通过燃烧反应,从燃料电池和普通电池中得到的能量受限于原子及分子本身的性质。对于一个特定的化学反应,你只能从中得到有限的化学能量。有没有办法可以得到更多更便宜的能量,甚至是一本万利呢?

答案是肯定的,那就是核反应。与其使用化学方法通过重组原子核和电子来获得新的分子,我们还不如通过重组原子核中的质子和中子来得到新的原子。通过核反应所得到的能量,可能要比通过化学反应得到的 100 万倍还多。但不幸的是,我们必须使用放射性同位素,并让它们以一种可预见的方式来衰变。铀、钚、钍等元素是最好的核燃料,但它们的储量稀少且价格昂贵。

当然,使用核反应也会给我们带来麻烦。一方面,铀、钚等既能用于发电,也可以用于制造核武器,造成核威胁。另一方面,大多数人并不愿住在核电站的周围,也不想生活在核废料堆的旁边,更害怕在核爆炸瞬间产生的巨大核辐射中枉死。

不可否认的是,无论是在当下还是在未来,核反应都会是工业化生产所需能量的重要来源。除了核裂变,核聚变以及其他一些更高能量的亚原子核反应也可以用来产生巨大的能量,虽然它们的可行性还有待证实。这张图片显示 π 介子(高能物理学家最感兴趣的人造粒子之一)与氢原子碰撞后的运动轨迹,这些轨迹是以氢气泡连接而成。而这些气泡则是在接近绝对零度的液态氢中产生的。

火　　焰

Flame

57

我们日常生活中所需的能量可分为热与功，热用于取暖与烹饪，而功则与运动或劳作相关。大部分(85%)所需的能量都是通过燃烧汽油、沼气、煤、木材等获得的。因为我们对火非常熟悉，所以燃烧看起来是如此的简单明了，但事实上并非如此。

从温暖的阳光到炽热的火焰，在这一系列的过程中都会涉及细小的结构。藻类和植物会通过自身复杂的光合作用结构来捕获阳光。它们凋亡后，会经历漫长的化学过程从而转变为石油，这一过程中需要热、压力以及粘土颗粒来辅助完成。而将石油转变为汽油则需要借助于大规模的工业化过程，并且要使用铂纳米颗粒来催化一系列的反应。取决于燃烧中的化学反应，不完全燃烧时产生的一些炽热的碳质小颗粒会使火焰呈现出不同颜色，而浓烟则是一些燃烧后仍然残留的碳质小颗粒。

那么火焰本身呢？不管是最小还是最大规模的，火焰都有着共同的特点。一支燃烧的蜡烛与燃煤发电并没有本质的区别，只是后者用煤生火并涉及柴油发动机。左图所展现的是蜡烛燃烧的情景。火焰散发的热量先将蜡烛熔化，使液态的蜡汇聚在烛芯处，再将其蒸发到空气中，蒸汽与空气混合，随后在火焰温度最高的区域发生一系列复杂的化学反应，最终将蜡和氧气转化为二氧化碳和水。而火焰的中间部分，则是一些未燃烧的碳质小颗粒，在接近1000摄氏度的高温下呈现黄色，当燃烧不完全时，剩余的碳冷却后形成烟或灰。

HEAT (to ward off the cold and cook food) and work (to help with the day's labor)—we have always needed both forms of energy. Most (85 percent) of the energy we humans now use comes from burning something—gasoline, methane, coal, wood. Burning seems straightforward, because flame is so familiar—but it is not.

Small structures occur everywhere along the chain of events that starts in the warmth of sunlight and ends in the heat of a flame. The capture of sunlight by algae and plants involves the complex biological apparatus of photosynthesis. The plants die; the very slow chemistry that later converts them into petroleum requires heat, pressure, and particles of clay. Transforming petroleum into gasoline involves industrial chemical processes carried out on an enormous scale, and catalyzed by nanoscale particles of platinum. A flame takes some of its color, and important parts of its chemistry, from small particles of incandescent carbon generated by incomplete combustion. "Smoke" is small particles of carbon remaining after combustion.

And the flame itself? The smallest flames share features in common with the largest: a burning candle tells the story as well as a coal-fired electrical power plant; only details are different in a coal fire and a diesel engine. Here, the heat from the flame melts the hydrocarbon candle wax; the liquid wax climbs up the wick; heat radiated from the flame vaporizes the wax; the vapor mixes with air; a complex series of chemical reactions in the hot region—the flame—convert wax and oxygen into carbon dioxide and water. At an intermediate point in the flame zone, small particles of unburned carbon—at a temperature of approximately 1,000 C—glow yellow. When combustion is incomplete, unburned carbon particles cool to smoke or soot.

Fuel Cell 燃料电池

58

MOST ENERGY IS GENERATED USING HEAT. Heat a gas, and it expands. Expanding gas can push a piston or spin a turbine.

Heat usually comes from burning something—coal, kerosene, natural gas, hydrogen. When a fuel—say, hydrogen (H_2, the simplest molecule; two protons and two electrons)—burns with oxygen (O_2), electrons move among atoms, the hydrogen and oxygen disintegrate and reform as water (H_2O), and the temperature soars. After hydrogen and oxygen have burned, very hot steam—water vapor—is all that remains.

There are few alternatives to hot, expanding gas as a way to get from fuel to electricity. One is a fuel cell. It also burns hydrogen and oxygen, but in a subtle, two-step way. A fuel cell is an elegant device that separates hydrogen from its electrons, uses these electrons to run a motor, and only then allows them to combine with oxygen and form water. The result is electrical power with little heat.

A fuel cell has two chambers, separated by a thin membrane. Hydrogen flows into one, air into the other. In the first chamber, on one side of the membrane, nanoscale particles of platinum separate electrons from hydrogen and send them into a wire that goes to an electrical motor. The protons that remain flow through the membrane into the second chamber. Here, on the other side of the membrane, similar nanoscale particles combine electrons leaving the motor, oxygen from air, and protons, and form water. The final result is the same as burning, but the procedure is different.

It sounds wonderful, and it can be. But like many wonderful things, fuel cells are expensive and complicated, and they sometimes drip. The present generation of fuel cells really only works well with hydrogen. And although there are many ways of generating hydrogen, they are all expensive, and the immediate benefits of hydrogen fuel cells—silent operation and clean air—may not justify their high costs.

This black film is the membrane, exposed for examination rather than buried in the interior of the cell. It is an astonishing piece of engineering. Protons (formed from hydrogen when its electrons are stripped away) flow though nanoscale pores in it; nothing (ideally; there is always some leakage) flows in the opposite direction other than water. The membrane's exquisite selectivity for protons is essential for the efficiency of the cell.

大多数能量都是通过热产生的。当气体受热时，它就会膨胀，而膨胀的气体则能推动活塞做功或使涡轮旋转。

热能通常来自某些物质的燃烧，比如煤、煤油、天然气、氢气等。以氢气(化学符号为 H_2，最简单的分子，由两个质子和两个电子组成)为例，当其在氧气(O_2)中燃烧时，电子在原子之间迁移，氢气和氧气分解后的原子重新结合生成水(H_2O)，并释放出热量使温度快速升高，热气腾腾的水蒸气是燃烧过后留下的唯一产物。

当然还有其他的方法来产生能量，其中最典型的就是燃料电池。燃料电池仍然是建立在氢气和氧气燃烧的基础上，但其原理和具体过程则完全不同。燃料电池是一个很巧妙的装置，它能将电子从氢原子中取出来使电动机运转，然后再让它们和氧结合生成水。如此一来，电池工作中产生的热量少了许多，取而代之的则是电能。

燃料电池有两个被薄膜隔开的容器，一个容器被注入氢气，另一个则被注入空气。在第一个容器里面，铂纳米粒子将电子从氢原子中取出来，并传送给与电动机相连的电线，而生成的质子则通过薄膜进入第二个容器内。在膜的另一边，类似的纳米粒子能使离开电动机的电子和空气里的氧以及穿过膜的质子相结合而生成水。虽然最终的结果与氢气燃烧是一样的，都是消耗氢气和氧气而产生水，但二者的过程却截然不同。

用燃料电池来产生电能听起来十分美妙，并已付于实践，但与许多美好的事物一样，燃料电池也有不少缺点，诸如价格昂贵、工艺复杂，有时候在使用过程中还会渗出液体。现有的燃料电池必须以氢气作为燃料。尽管我们有很多方法来产生氢气，但它们的成本还都太昂贵。到目前为止，虽然氢燃料电池有着低噪声和无污染等优势，但远不足以弥补它高成本的缺点。

图中展示了一张黑色的薄膜，它就是燃料电池中那张神奇的膜，它只允许质子通过其内部的纳米孔道，除了水没有其他任何物质能从其相反方向流过来。使用对质子有高选择性的膜是制备高效率燃料电池的必要条件。膜的质量好坏直接决定了燃料电池的工作效率。

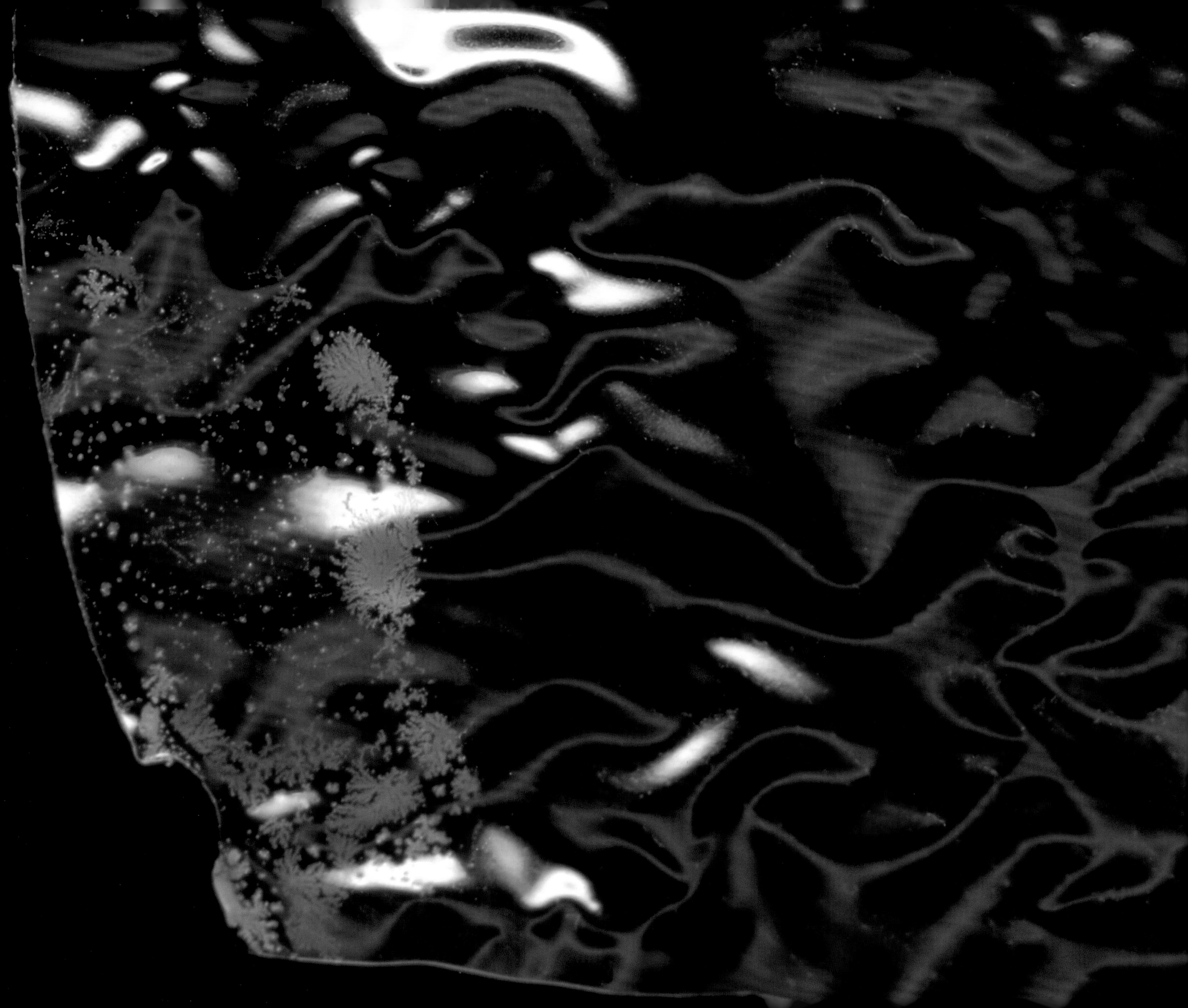

太阳能电池

Solar Cell

59

不得不承认，人类长期以来已养成了坐吃山空的习惯，并喜欢不劳而获。我们通过燃烧"死亡的植物（如煤和石油）"来获取能源。然而，这种方式已经远远满足不了日益增长的需求。尽管地球上有足够多的煤让我们在很长一段时间内维持温暖(石油和天然气可能够呛)，但前景仍不容乐观。燃烧产生的热量会造成海平面上升，其产生的二氧化碳也会让地球的气候变得越来越糟，最终像金星上那样不舒适。千万不要幻想我们可以向北极搬迁以躲避不断上升的海平面，就算你能跟北极的鳄类动物和平共处，这个想法也如南柯一梦般不切实际。作为新的途径，我们开始尝试使用核能以及风能，并渐渐尝到了甜头，但它们也有自身的问题。为什么不考虑一下太阳能呢？

为了产生电，太阳光必须照射到一块充满电子、薄薄的硅板上（实际上它是由性质稍微有些不同的两块板面对面构成的）。有时候吸收一个光子会使硅板上的电子偏离其原来的位置几百纳米，甚至越过两块板之间的界面到达另一块硅板。如果一切顺利（当然，通常没这么简单），游离出去的电子会通过外部的电线回到它们原来的位置，这一过程就产生了电，可用来驱动电动机或生产氢气。太阳光是免费的，硅太阳能电池也能用上好几载。还有什么比这更完美的呢？

任何事物都存在着两面性，有优点必然也有缺点。太阳光是免费的，但是收集太阳光却需要耗费大量的人力和财力。硅太阳能电池的价格相当昂贵；而且，在太阳落山后，我们还得继续生活，也许太阳能可以储存在蓄电池内，供晚上使用，这又需要额外的投资。另外，灰尘、落叶、雪花以及鸟粪也都会减弱太阳能电池对光的吸收。

太阳能发电是一个非常好的想法，但它还没有好到能够完全满足人类活动所需的程度。为此，我们还需要省着点或继续寻找其他能源。

WE LOVE THE IDEA OF something for nothing. We acknowledge, grumpily, that eventually our habit of picking over the graveyards of dead plants—coal deposits, oil fields—and burning what we find to generate energy will not keep pace with our appetites. Although we have enough coal to keep us warm for a long time, oil and gas are less certain, and anyway there's the nasty prospect that the carbon dioxide we produce will edge Earth's climate uncomfortably far toward that of Venus. Maybe we can tolerate crocodiles at the North Pole, but moving most of our civilization to avoid the rising sea? Bad idea. Alternatives to fossil fuels—nuclear power and wind—have both disadvantages and advantages. So why not solar power?

To generate electricity, photons from the sun—sunlight—must fall into the sea of electrons in a thin plate of silicon (actually, two face-to-face plates, with slightly different properties). Absorption of a photon sometimes kicks an electron a few hundred nanometers from its customary place, across the boundary between the plates. If everything works (and it usually doesn't), the itinerant electron may return home by going through an external wire, and turning a motor or generating hydrogen along the way. Sunlight is free; silicon solar cells operate for decades; they have no moving parts.

But there's always a catch. Sunlight is free, but *collecting* it is not. Silicon solar cells are expensive. And much of our life goes on when the sun is down: what about the batteries to provide electricity at night? Dust, leaves, snow, and bird droppings all dim the vision of solar cells.

Solar electricity is a good idea, but not a good enough idea to save us from ourselves. Either we have to find more energy elsewhere, or use less.

Plants and Photosynthesis

植物和光合作用

60

FROM SPACE, EARTH IS BLUE, white, green, and yellow. Blue for ocean, white for cloud, green for plants, and yellow for desert. Life depends on sunlight, water, and plants. We humans are almost unnoticeable.

Plants use sunlight, water, and carbon dioxide to make more plants. The machinery of this process—photosynthesis—is highly optimized: plants and their forebears have been perfecting it for several billion years. Plants were here long before we were, and we are here—eating them, and breathing the oxygen they exhale—as a sideshow in their photosynthetic circus.

The leaf is where the main events of this circus occur—"The Capture of the Photons! The Reduction of Carbon Dioxide!" The green color of leaves is the light they reject: they absorb mostly red. (The regions of red in these leaves show that there *are* other absorbers, with other colors.) The energy picked up from sunlight converts carbon dioxide to sugar (glucose) and incidentally generates oxygen. Plants build themselves out of glucose (as cellulose) and discard the oxygen. We need both: their glucose (as starch) is our daily bread, and their oxygen is the air we breathe. Plants make it; we mammals burn it.

Most of the energy we use—for heat, light, transportation, and the work of the world—also comes from burning the products of past photosynthesis, either directly (wood, peat) or after millennia of further transformation (coal, petroleum, natural gas). And now we expect plants to take up the carbon dioxide we produce by burning the remains of their ancestors, and reconvert it to something useful (or at least get it out of the way).

Most of our energy, all the oxygen we breathe, all the food we eat, the repair and replenishment of our atmosphere—that's a lot to ask of plants! We are less than grateful in return.

从太空上望下来，地球呈现五颜六色，有蔚蓝的海洋、洁白的云朵、墨绿的植被以及金黄的沙漠。人类能够在地球上生存，得益于阳光、水和植物。相比之下，人类显得有点微不足道。

植物利用阳光、水和二氧化碳来繁殖，这一过程的根本机制就是光合作用。植物以及它们的祖先已经用了几十亿年的时间来进化和完善这种机制，使其高度优化。很久很久以前，植物就已在大地上生存。它们不仅孕育了人类，还将自己作为食物奉献给人类，并提供氧气给人类呼吸。我们在植物光合作用这个魔术表演中仅仅是一个小小的配角。

叶子是这场表演的主角。“捕获光子！还原二氧化碳！”因为大多数叶子主要吸收红色光，所以它们通常呈现出白光中没有被吸收的光的颜色，即绿色(图片中叶子上红色的区域说明它里面还有能吸收其他颜色光的分子)。叶子借助太阳光的能量将二氧化碳转化成糖类，如葡萄糖，并附带产生氧气。植物借助这些以纤维素形式存在的葡萄糖得以生长，并释放氧气。而这两种物质恰是人类生活的必需品：植物产生的、以淀粉形式存在的葡萄糖是我们每天的主食，而氧气更是我们片刻也不能脱离的。植物生产它们，我们则消耗它们。

无论是直接燃烧木材或碳，还是经过几千年转化而成的煤、石油和天然气，我们从自然界所获取的能源大多数都来自植物的光合作用，这些能源也被广泛地应用在我们的日常生活和工作中，例如加热、照明和运输。在使用这些能源的过程中，我们又产生了大量的二氧化碳。如今，我们反过来期望植物能清除这些通过燃烧和消耗它们祖先的遗体所产生的二氧化碳，甚至将其转化成一些有价值的东西。

人类赖以生存的大部分必需品，包括呼吸的氧气、吃的食物、洁净的环境，都来自植物的馈赠！然而，我们在感恩回报方面做得还很欠缺。

CODA

We glance, and turn away without noticing. We don't ever really see, and then we forget what we have seen. Water drips from faucets; candles burn; yeast makes bread rise; a tiny, living mouse—pursuing its tiny murine intentions—runs across a floor that was once a living tree; the sun consumes itself. We don't notice.

Look more closely, and everyday events bloom into a reality so transfixingly marvelous that you can't look away. Life becomes something we don't understand that happens in ordinary matter. Ordinary matter happens somehow when atoms get together. Atoms build themselves from electrons and nuclei, following rules that flummox intuition. Electrons and nuclei are strange avatars of yet stranger fish swimming in a darker sea.

But whatever it all is—this amazing assembly we so flippantly nickname "reality"—is all there is, and all we are.

尾声

我们匆匆一瞥,然后漫不经心地把脸转过来。因为从未用心去看,以致于我们很快就忘记了曾经目击到了什么。太阳在自我消耗、龙头在滴水、蜡烛在燃烧、面在发酵膨胀……为达到自己小小的目的,一只活蹦乱跳的小老鼠在地板上窜来窜去。而那地板,曾经也是有生命的树木。所有这一切,我们都会视而不见。

看得再仔细一点,每天所发生的事情构成了我们丰富多彩的生活,常常令人难以忘怀。生活中的任何物体都由原子组成。而原子又是由原子核和电子构成,仅仅是不同的排列就能让世界变得丰富多彩,真让我们百思不得其解,更不用说去理解生命的真谛。

不管这些故事听起来多么匪夷所思,这五彩缤纷的世界都是客观存在的,包括我们自己在内。

FIVE NOT-SO-EASY PIECES

Notes from the Photographer

FROM THE BEGINNING, GEORGE WHITESIDES AND I understood that many of the images in *No Small Matter* would, of necessity, be metaphoric in order to communicate the invisible complexity of the micro and nano worlds. He started a list, which eventually expanded with various potential concepts coming from both of us. To suggest the nature of reality, George at first proposed a salami sandwich ("or pick your spread—peanut butter would be fine, but the filling has to show, and it has to be thin"). At that moment I panicked, thinking I was in deep trouble—I'm not good at salami. But over time we worked our ideas back and forth until we were both satisfied with the representation, scientifically and aesthetically. Our two perspectives became complementary. I presented him with a variety of photographic ideas that seemed to capture key concepts, and nudged him to think and write metaphorically. And his creative explanations of the science encouraged me not to be content with rounding up the usual suspects, or with just another pretty picture, but to aim for something more expressive.

五件不那么容易完成的作品

(来自摄影师的注释)

从一开始,George Whitesides 和我就都明白大多数用在《见微知著:纳米科学》中的插图必须有隐喻的含义,这样才可以用来描述无影且复杂的纳米和微米世界,从而更容易地向读者传达准确的信息。

他给了我一个清单,上面包含了各种概念在生活中的影射,其中一个是意大利香肠三明治(什么样的面包都可以,也可以加点花生酱,但是里面的内容要薄,同时也要能显露出来),他觉得这是一个描述"现实"生活的好例子。当看到这个提案后,我感到很为难,因为我根本连怎么切香肠都不知道。后来,我们反复交流了各自的想法,尽量让图片既符合科学又引人入胜,直到大家都满意为止。我们从两个不同的角度实现了这本书的互补性。我向他展示了对各种图案的想法,让人通过看图就能捕捉到关键的概念,同时我还说服他从图片出发用隐喻的方式去思考和写作;他对于科学概念的创造性解释则鼓励我去选择更具表现力的图片来阐明问题,而不仅仅只是追求美观。

QUANTUM APPLE

量子苹果

To ILLUSTRATE the counter-intuitiveness of the quantum world, I started with an image of a modified mosque and ended up with an image of a glass apple. How did this happen? In the first image, I was attempting to create a place that could never exist, by digitally combining what I thought were two incompatible buildings: the Salk Institute in La Jolla and the Hagia Sophia in Istanbul (both taken in my former life as an architectural photographer). But after some of my colleagues called the new "place" beautiful, I knew that the image wasn't suggesting the idea of uneasy contradiction that I had in mind. So I dumped it and moved on.

I had always loved a glass apple that my late husband received as a teaching award in surgery. It is both visually and tactilely pleasing, and seemed in some way to crystallize the notion of a quantum world where an apple isn't really an apple anymore. I photographed it from

为了描述有悖于直觉的量子世界，我最初打算使用一张经过处理的建筑物图片，但最终却选择了一个玻璃苹果的图片。为什么做这样的改变呢？在原来的图片中，我本想用数字化的手段将两栋互不相容的建筑，即位于拉荷雅的索尔克研究所和位于伊斯坦布尔的圣索菲亚大教堂相结合（这两张照片都是我以前作为建筑摄影师时拍摄的作品）。但是，当同事们看到这幅图片后都赞美这个"新建筑"很漂亮时，我才意识到这张图片不能表达我脑海中所想象的那种视觉上的冲击。因此，我放弃了这个方案，去寻找更合适的。

这个玻璃苹果一直是我的至爱，因为它是我已故丈夫在讲授外科手术课时获得的奖品。不论在视觉上还是在触觉上它都让人赏心悦目，并且可以在一定程度上将量子世界的概念具体化，即这个苹果在量子世界里已不再是一个苹果了。我从苹果的上方拍摄它，但结果却不那么令人满意。随后，我又尝试从正面的角度去拍摄，并在后期图像处理时把某些发散的反射光删掉，这样效果就好多了。

接下来的挑战是如何在没有突兀感的情况下启发读者产生矛盾的意识。我决定用阴影来解决这个问题。我找到一个金属的立方体并让它投射出与自己形状相符的影子，再用数字化的手段把这个阴影粘贴到那张苹果的图片上，然后进行横向与纵向的拉伸、扭曲等操作，以唤起读者有违直觉的感受。我还一度在阴影中剪了一个方形的洞，但是这样的表达好像稍微有些夸张，所以后来我还是选用了完整的矩形阴影来表达这一思想。

an overhead angle, and when this was unsatisfactory, I tried a more frontal point of view—much better, although a distracting reflection from one of the light sources had to be digitally deleted.

The next challenge was how to suggest the idea of contradiction without screaming it at the reader. The solution, I decided, was in the shadow. The only thing I could find at the time to cast the shape of shadow I needed was a metal cube. I digitally pasted just the cube's shadow onto the apple photograph, and then stretched, pulled, and distorted it, trying to evoke a sense of counter-intuition. At one point I even cut a square hole in the shadow—a bit of overkill, I admit. After that, I decided that a simpler rectangle of shadow would work just fine.

COUNTING ON TWO FINGERS 二进制计数

THIS WAS ONE of the first images I began to think about for this book. At first, my understanding of exactly what "binary" means was not clear. I thought I needed to show not 0's and 1's (nothing and something) but two states (this and that). So I started with a random pattern of colored glass pebbles, with no intended numerical meaning. In a second try, I used a more identifiable material, although few readers would guess that what they see on the left is salt and pepper.

George was concerned about the regularity of these images and suggested a pattern that could be translated into a binary equation. And to capture the idea of electron flow, he asked me to use a liquid that is present or not present in some sort of vessel. I began with water in one of my favorite sets of glasses and then tweaked the image in Photoshop. But when trusted friends told me the result was difficult to "read," I shifted the camera to an overview.

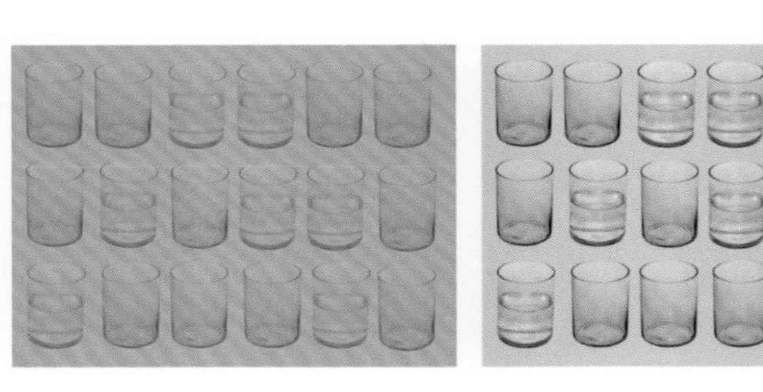

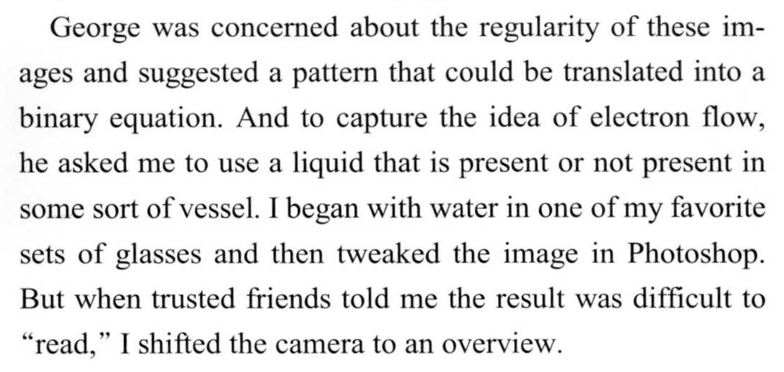

这是一张我最初想要放入这本书里的图片。当时我对"二进制"的准确含义理解得并不够透彻。我以为我需要展示的不是0和1(无和有),而是两种不同的状态(这个和那个)。所以,我起先将绿色和橙色的玻璃球随意地摆放,没有包含任何数字的含义。在第二次尝试中,我使用了更容易区分的材料:盐和胡椒。

George非常关心图案的有序性,并建议我将图案中的各个元素摆放成二进制等式。为了表达电子流动,他让我在某些容器中盛装液体,另一些则空着。于是我挑选了自己最喜欢的一组玻璃杯,装上水拍照,然后用Photoshop软件进一步处理。当朋友们告诉我这张图片很难看懂时,我又试着从杯子顶上直拍下来。

在最后一次试镜中，我将Shiraz葡萄酒装在一些厚壁玻璃杯中，然后将它们与空杯子组合在一起拍摄。这样拍出的效果令人相当满意。为了庆祝这个得意之作，拍摄结束后我喝光了杯中所有的葡萄酒。

	上面一排	0 0 1 1 0 0
+	中间一排	0 1 0 1 1 0
=	下面一排	1 0 0 0 1 0

代表着

$$12 + 22 = 34$$

In what turned out to be my final shot, I filled a heavier glass with a wonderful Shiraz I was eager to try and photographed it, along with another unfilled glass. The result was quite satisfying (as was the glass of wine, which I emptied in celebration).

	Top row	0 0 1 1 0 0
+	Second row	0 1 0 1 1 0
=	Bottom row	1 0 0 0 1 0

Representing $12 + 22 = 34$

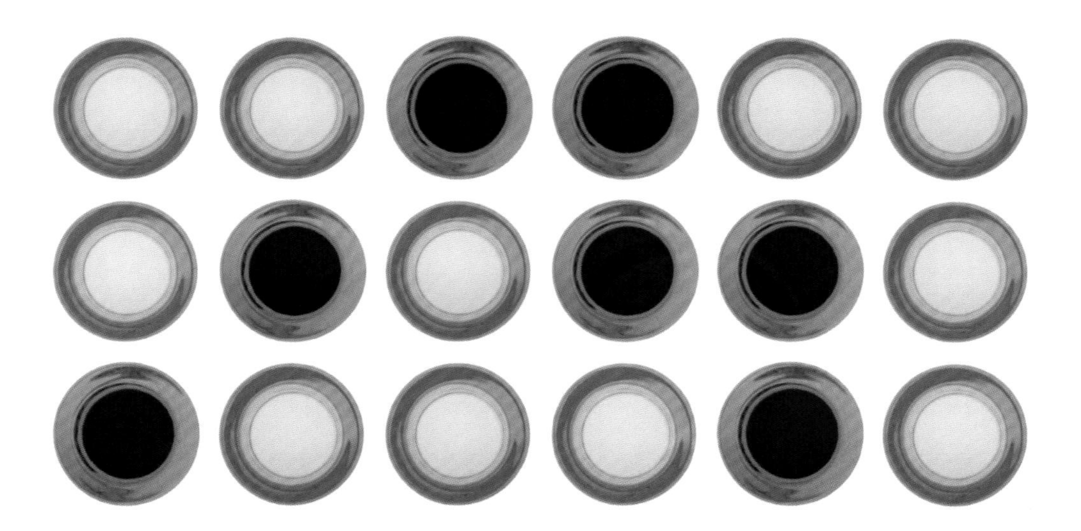

NANOTUBES

纳米管

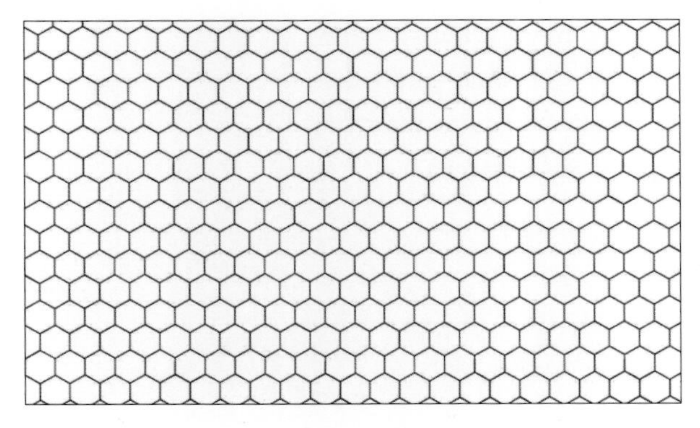

WHEN WE DECIDED to include an image of nanotubes, I approached the project photographically, as I usually do. But in my mind, the obvious image—a scanning electron micrograph—was not an option, because others have already done that, and much better than I probably could. The most productive course, I decided, would be to simulate nanotubes.

First I printed a black hexagonal pattern, representing a standard graphite lattice, on an 8×10 piece of transparent acetate. As I began to roll the acetate to make a tube, I had to decide how to connect the edges of the paper. Scientific sources informed me that there were indeed various possible configurations for carbon nanotubes, and that the ultimate configuration was significant in determining the electrical properties of the structure.

I decided (for aesthetic reasons) to adopt what's called the "zigzag" configuration. I secured the edges of the acetate with a couple of pieces of tape and placed the tube on my flatbed scanner.

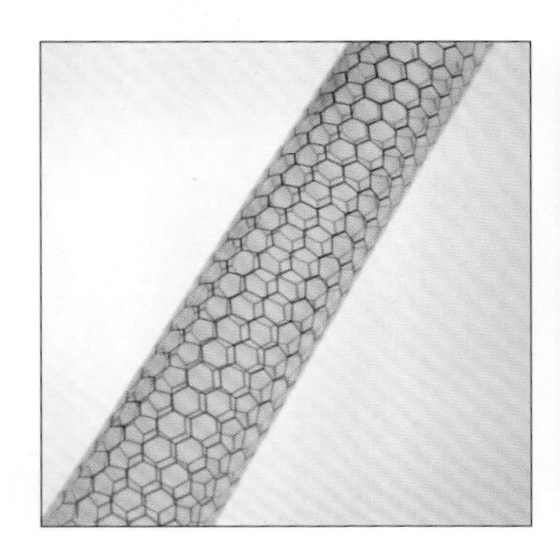

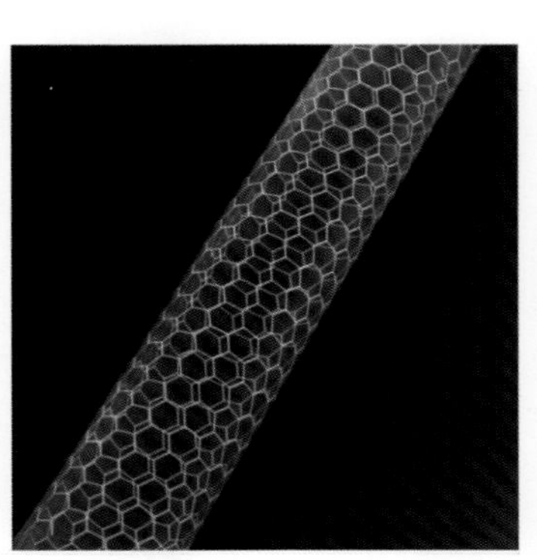

当决定在书中展示一张纳米管的图片时，我像往常一样打算用摄影的方式来完成。在我看来，使用纳米管的透射电子显微照片并不是很好的选择，因为有人早已这样做过了，而且很可能比我拍摄得更好。我觉得用仿真模拟的方式来制作纳米管的图片会更加有效。

首先，我在一张 8×10 的胶片纸上打印出黑色六边形的图案来代表标准的石墨烯晶格。经验告诉我，要想将这张打印好的膜卷成管，就必须要把两条边缘完美地连接起来。因为碳纳米管有着各种各样的构型，采取哪种卷曲方式将决定它的电学性质。

从美学的角度考虑，我决定采用“之”字形。我用胶带将卷曲的胶片纸的边缘牢牢固定，并把它放在扫描仪上，这样扫描出的结果不是很吸引人。于是我又用Photoshop软件将纳米管反转，再把这些图片复制几份后叠加在一起，通过调节图层的透明度，就得到了多层纳米管。最后，我又根据感觉对图片进行了滤镜及反转等处理。

研究和创作这张图片的过程让我学到了很多关于纳米管的科学知识。无论你是正在暗室里处理图片的摄影师、在黑板上画图的物理学家，还是正在作曲线函数图形的学生，都是在利用图片来了解事物本身。

The result was not terribly compelling. I then “inverted” the nanotube in Adobe Photoshop. Going further, I combined a few replications of the image to make multiple layers with varying degrees of transparency. Then, for what I thought would be the final composite, I adjusted the image using various filters and additional inversions.

The act of researching and creating this image taught me quite a lot about nanotube science. Whoever you are—a photographer manipulating a scientific image in today’s virtual darkroom, a physicist sketching on a blackboard, a student reaching for a visual metaphor or graphing a function—understanding often comes through the act of representation.

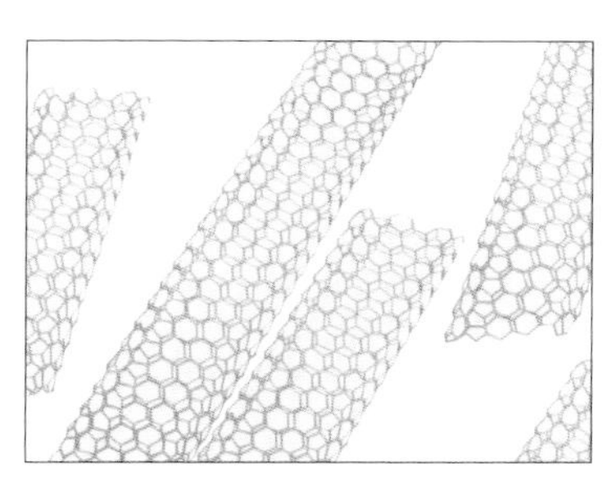

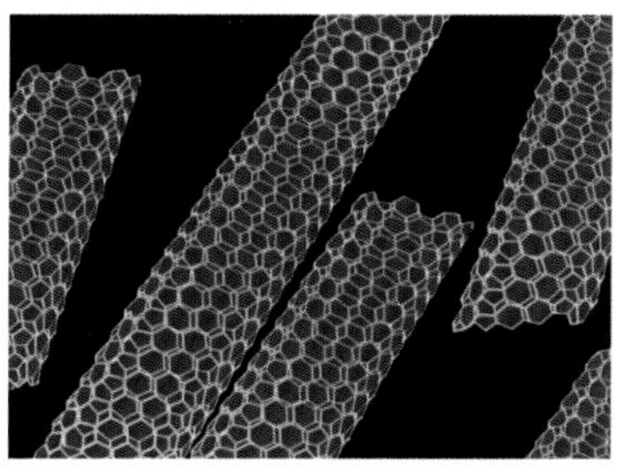

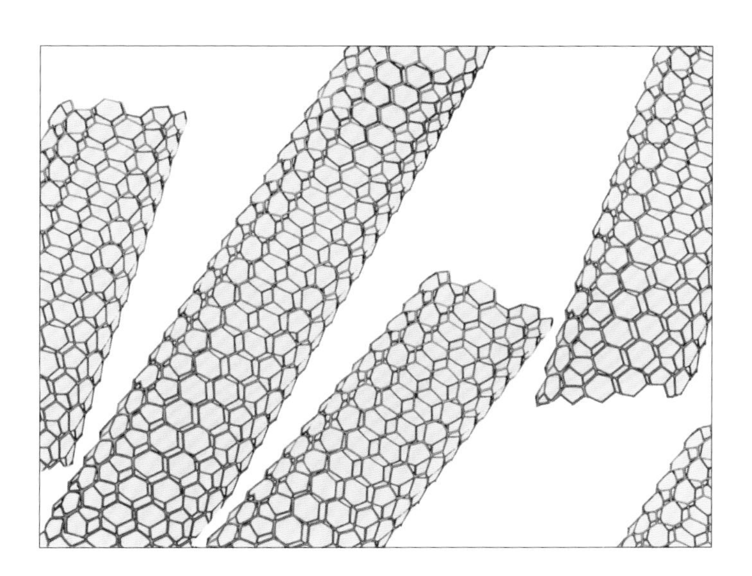

PRISM AND REFRACTION　　棱镜和折射

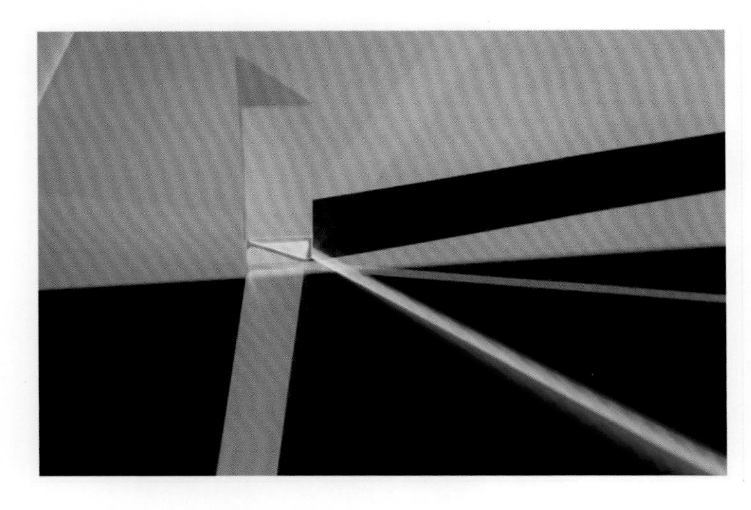

WHEN George and I began our list of possible images, he wrote to me that "most of quantum mechanics turns out to be represented in interference and diffraction." He went on to suggest a few ideas for that concept, including a standard-issue photograph of light dispersed within a prism. So using a glass prism I purchased on eBay (which came in a "handsome, black, soft flannel travel bag"), I tried a number of garden-variety photographs, some with a green laser pointer along with varied light from a window.

None were satisfying, even after cropping. For days, I walked around my house holding the prism and seeing some pretty spectacular spectra displayed on walls, ceilings, sofas, and all the rest. But again, nothing that I hadn't seen before in a text-

当 George 和我在清单上列出可能要放进这本书里的图片时，他写了这样一句话："量子力学基本上就是干涉和衍射。"他还给了我一些建议，包括一张经过棱镜分光的标准图片。因此，我从网上买了一个玻璃棱镜，它被包装在一个漂亮的黑色绒布袋里，每一个镜面都是那么的平整和光滑。我用了很多方式来拍摄分光图案，包括用绿色的激光笔和窗户投射来的自然光来照射它。

但是，即使是经过后期处理，这些照片仍没有一张令我满意。接下来的几天里，我都拿着这个棱镜在家里漫步，看看棱镜在墙上、天花板、沙发或是其他一些东西上是否能投射出更漂亮的图案。结果同样令人失望，我所看到的图案基本上跟教科书或网络上找到的差不多。

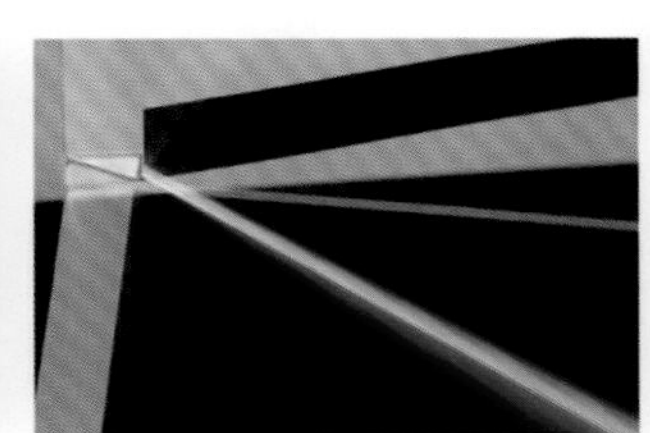
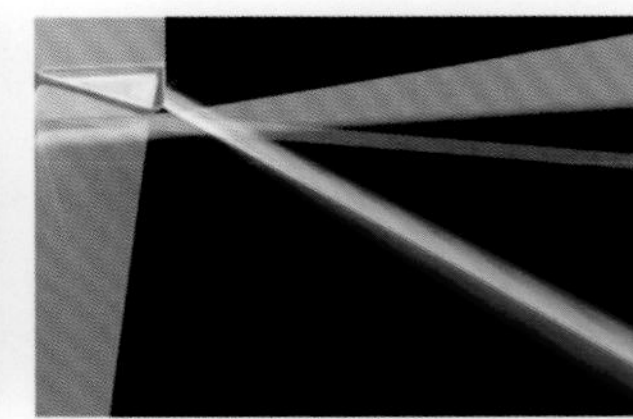

直到有一天，外面乌云密布，我把棱镜放在一对皮制盒子上，然后把它们固定在一个很大的百叶窗前。百叶窗产生的白色与黑色图案，以及光线在棱镜内部形成的五彩斑斓的颜色，让整张图片充满了强烈的对比度。这个结果让我欣喜若狂。

我将拍摄到的图片旋转了180度，然后放进了这本书里。这样做的目的仅仅是为了让它看起来更加美妙。

book or blog jumped out at me.

Finally, on a cloudy day, I found what I was looking for. I propped the prism on a couple of leather boxes (which happened to have stitching) in front of venetian blinds drawn on one of the larger windows. The contrast between the black and white pattern of the blinds and the brilliant colored lines inside the prism was dramatic and surprising.

For the book, I rotated the image 180 degrees. It just looked better.

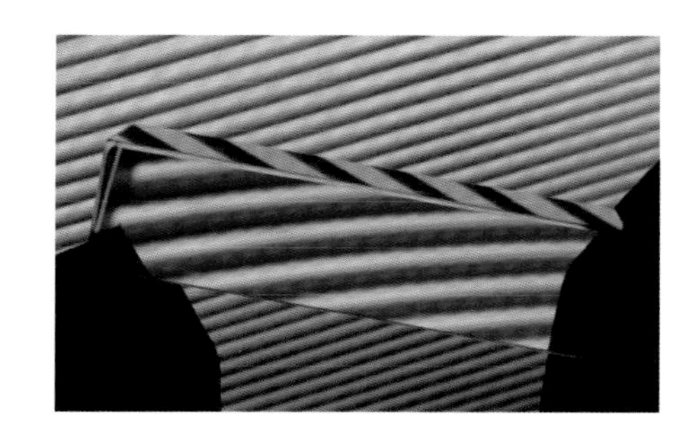

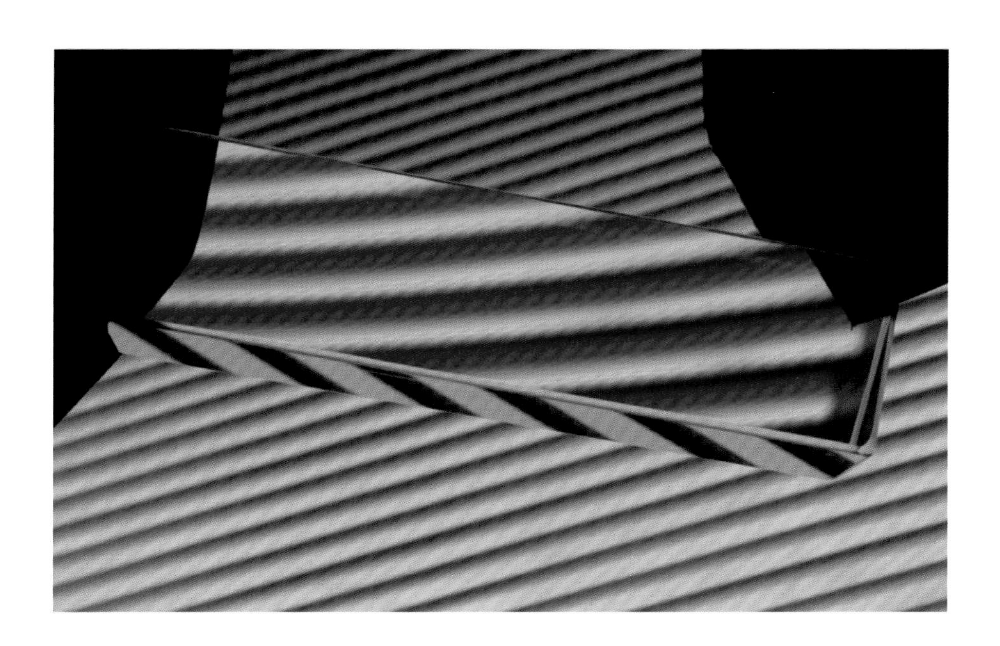

QUANTUM DOTS AND THE CELL

量子点和细胞

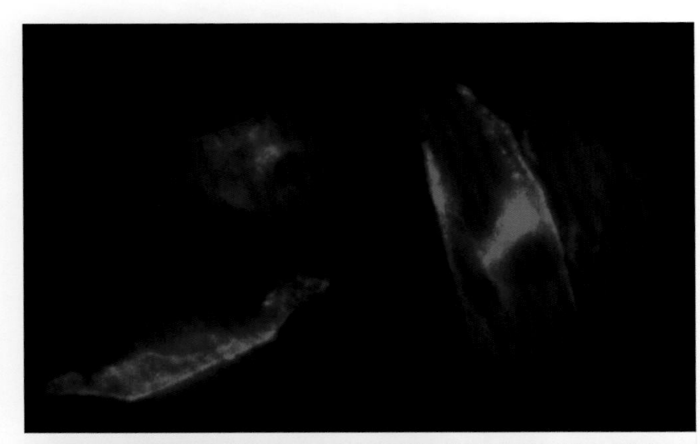

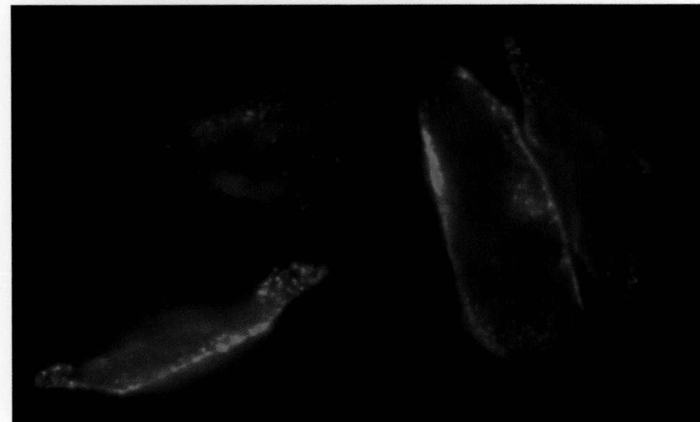

ALICE TING AND Peng Zou supplied me with two of their original images. The colored areas give highly detailed morphological information about the cell, as specific quantum dots attach to specific structures. When the dots are excited with ultraviolet light, they fluoresce and so we see where they attach.

In standard views of fluorescently labeled material, the colored areas are set against a black background. I thought it would be interesting, and perhaps informative, to see the information differently, as long as I maintained the integrity of the science. So I "inverted" the two images in Photoshop to get a white background. But because it was important to keep the original colors for the structures themselves, I changed the colors back to the red and green.

Alice Ting 和 Peng Zou 为我提供了两张他们拍摄的原始荧光显微图片。其中，表面经过特定修饰的量子点会选择性地吸附在一些特定的细胞器上，当量子点被紫外光激发后，便会发出某种特定颜色的荧光。借助荧光显微镜我们就能看到它们在细胞中所处的位置。

在通常的荧光显微照片中，除了量子点标记的区域，其他地方都是黑色的。我觉得只要能保持结果的科学性和真实性，对图片进行一些处理应该会非常有趣，而且还有可能会给我们提供更多有用的信息。因此，我用 Photoshop 软件将这两张图片进行了"反相"处理，得到了白色的背景。但为了保持细胞结构的原始颜色，我又将反相后的有色区域改回成原来的颜色。

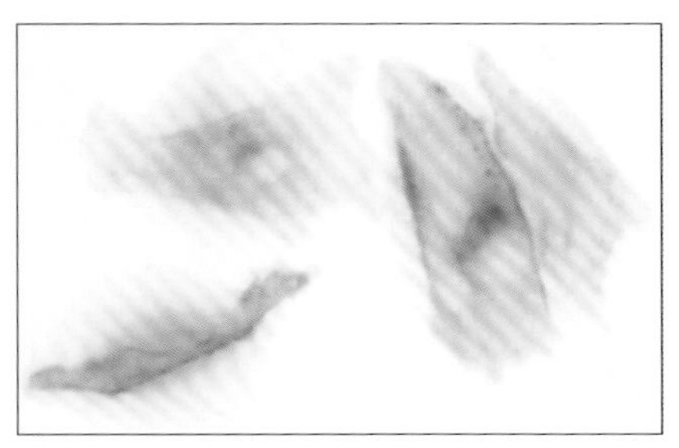

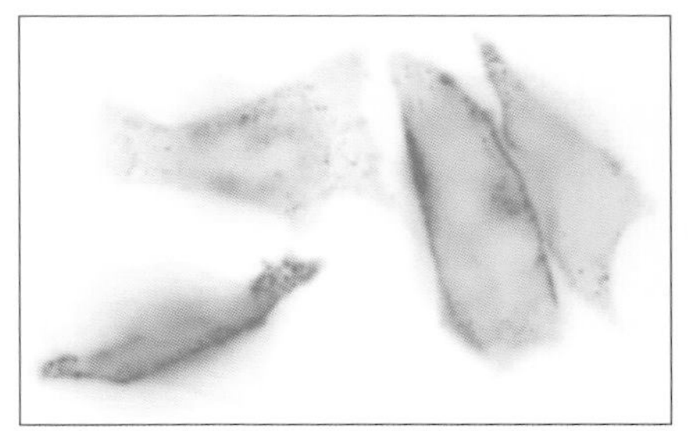

最后，我将这两张图片叠加在另外一张由常规显微镜在同一区域拍摄的照片上。显而易见，三张图片叠加后的结果仍然显示了与原始图片相同的内容，却大大增强了细胞的立体感。

I then layered both of the images over a third image taken with standard microscopy, again supplied by the laboratory. The three-layered result shows the same information as the original separate images, with the addition of a sense of the whole.

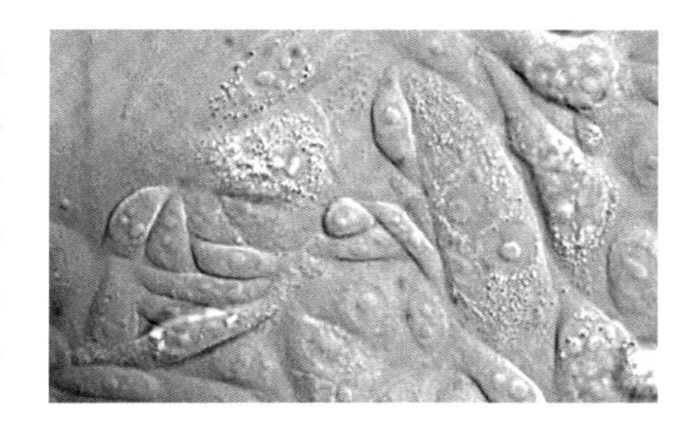

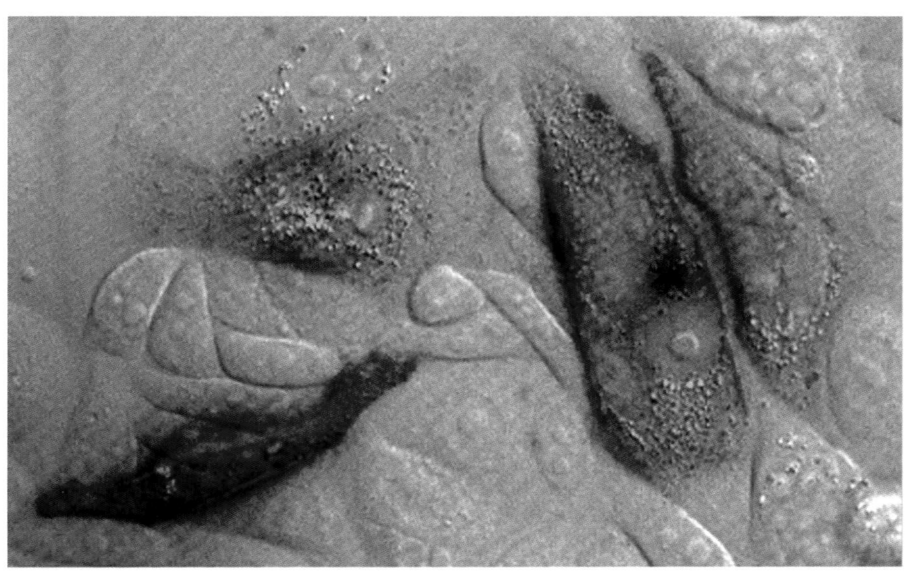

扩展阅读 Further Reading

Dyson, Freeman. *Imagined Worlds*. Harvard University Press, 1997.

Feynman, Richard P., Anthony Hey, Tony Hey, and Robin W. Allen. *Feynman Lectures in Computation*. Westview Press, 2000.

Ford, Kenneth W. *The Quantum World*: *Quantum Physics for Everyone*. Harvard University Press, 2004.

Gribbin, John. *In Search of Schroedinger's Cat*: *Quantum Physics and Reality*. Bantam Books, 1984.

Hey, Tony, and Patrick Walters. *The New Quantum Universe*. Cambridge University Press, 2003.

Hillis, Danny. *The Pattern on the Stone*: *The Simple Ideas That Make Computers Work*. Science Masters, 1998.

Laughlin, Robert B. A *Different Universe*: *Reinventing Physics from the Bottom Down*. Basic Books, 2005.

Ozin, Geoff A., Andre C. Arsenault, and Ludovicho Cademartiri. *Nanochemistry*: *A Chemical Approach to Nanomaterials*. Springer-Verlag, 2008.

Ratner, Mark A., and Daniel Ratner. *Nanotechnology: A Gentle Introduction to the Next Big Idea*. Prentice Hall, 2002.

Von Baeyer, Hans Christian. *Information*: *The New Language of Science*. Harvard University Press, 2003.

——. *Taming the Atom*: *The Emergence of the Visible Microworld*. Dover Publications, 1992.

跋

Acknowledgments

纳米科学的覆盖面十分广泛，我们通常被许多同行们引领着去不断思考这门学科。当然，我们首先要感谢对方，这本书反映和折射出我们之间对于纳米科学、科学以及社会的讨论，以及怎样帮助对方去理解一个具体的问题。虽然有不同观点或看法是一件有趣的事(尽管这也不那么容易做到)，但我们更希望把不同观点及看法融合在一起来描绘出一幅完整的图画。

除此之外，我们还分别对不同圈子的人有感谢的话要说，因为这些图片和文字都是通过各自的途径产生的，即使它们的归宿都是一样的。

来自 George M. Whitesides(原书作者之一)的感言：Mara Prentiss 教授毫不吝惜她的时间和耐心。多年来，她和我一起讲授一门关于纳米科学与技术的公共教育课。她对于细小物体的物理总会有新奇、意想不到和有趣的见解。Mara 真正懂得万物是如何工作的，包括我最感兴趣的一些问题。我对于微流控的大部分理解都是 Howard Stone 教授教给我的。DSRC(防御科学研究委员会，它是一个拥有科学家和工程师的小团体，他们建议 DARPA，即防御先进课题研究机构，致力于材料科学和纳米科学的研究)的成员让我对凝聚态物理中几乎所有令人感兴趣的问题有了一个初步了解。

大多数做研究的科学家都会讲同一句话：他们最好的老师通常是他们的学生。对于我来说，在哈佛大学与我研究小组里的研究生和博士后们一起工作真是一件最引人入胜的事。聪明的学生们对我及他们自己提出的问题常常令人兴奋不已，无论是知道还是不知道答案。我们小组关于纳米科学的研究极大地得益于与 Ralph Nuzzo(关于自组装单分子膜，现在被广泛地称作SAMs)和 Mark Wrighton(关于表面科学和软刻技术)的合作。Colin Bain 在 SAMs 的起始阶段作出了非常重要的贡献，Abe Stroock 和 Rustem Ismagilov 让微流控方面的研究逐步走上正轨。如果没有与 Don Ingber 长期和富有成效的合作，我们根本不可能开展在细胞生物学方面的工作。与比你聪慧和拥有更丰富知识的人进行合作是在科学研究中最让人兴奋的事情之一。

一份作者认为是“最终的”文本草案，往往只是编

NANOSCIENCE IS A VERY BROAD SUBJECT, and we have been guided by many colleagues in thinking about it. Our principal indebtedness is, of course, to each other: this book reflects and refracts years of conversations and arguments about nanoscience, and science, and society, and how to explain one to the other. It is great fun (although not always effortless) to see in different colors of the spectrum, but to try to paint a picture together.

Beyond that debt, we have different networks of indebtedness to others, since the development of image and text follow separate paths, albeit one linked by common endpoints.

From GMW: Professor Mara Prentiss has been uniquely generous with her time and patience. I have taught a general education course on nanoscience and nanotechnology with Mara for years. Her discussions of the physics of small things have *always* brought up something new, unexpected, and interesting. Mara really understands how stuff works, including some of the parts of the universe I'm most interested in. Professor Howard Stone taught me most of what I know about microfluidics. The members of the DSRC (the Defense Science Research Council, a small group of scientists and engineers who advise DARPA—the Defense Advanced Projects Research Agency—on topics that include materials science and nanoscience) have given me at least a familiarity with almost everything interesting in condensed matter science.

Most academic scientists will say the same thing: their best teachers are often their students. Working with the graduate students and postdoctoral fellows in my research group at Harvard has been the most engaging thing about a career in science. The questions smart students ask—of themselves, and of me—have made it impossible to be anything other than excited by what science knows, and even more by what it doesn't know. Research in our group in nanoscience has benefitted enormously from collaborations with Ralph Nuzzo (on self-assembled monolayers, now affectionately and widely known as SAMs) and Mark Wrighton (on surface science and soft lithography). Colin Bain was especially important in starting the work on SAMs, and Abe Stroock and Rustem Ismagilov in taking up the subject of microfluidics. Without our long and productive collaboration with Don Ingber, we could never have worked effectively in cell biology. Collaboration with people both smarter and more knowledgeable than you is one of the best things that can possibly happen in science.

A "final" draft of text (from the vantage of a writer) is usually the point at which the editors go to work to make something of it. The most important editor for the text in this book was my wife, Barbara Whitesides, whose detailed attentions enormously improved every aspect of the writing, from choice of metaphor to details of grammar; her skill in delicately rearranging words, and in pointing out ideas that I thought were good but that weren't, and in asking what on earth I was thinking of when I wrote *that*, have pulled large parts of the prose from something I could not bear reading to something that is tolerable. Susan Wallace Boehmer was the editor at Harvard University Press. She was essential to making the book clear, to refereeing differences of opinion, and to getting it finished. She was also important in wringing the excesses out of some of the metaphors, in restraining my instinctive improprieties, and in defending the prose. Becky Ward distilled simplicity out of complexity efficiently, gracefully, and with great skill.

One son, Ben Whitesides, with his lovely ear for prose, added a number of memorable phrases; the other, George T. Whitesides, conspired with Ben over many years to try to build a functional father of me. Being a father may seem unrelated to writing about science, but it is not. Among the many reasons that being a parent is the most wonderful of jobs is that it requires you to explain everything in the world in random order to curious children. If you can't first explain to yourself, you certainly can't explain to others. George and Ben, and Barbara, and Felice, Mara, and my students, have been my teachers, and I am—as a student should be—most grateful to them.

There are also a number of people—scientists and others—whose contributions to "nano" generally, and to my education in it, deserve specific thanks. Heine Rohrer and Gert Binnig invented the scanning probe microscope, and started it all; Cal Quate explained nanoscience to me in the beginning, and helped me to understand some of its opportunities and problems. Hank Smith taught me to think about nanolithography and nanoelectronics realistically. Mike Rocco was a tireless and effective advocate for the field in Washington in its infancy, and gave nanoscience a vital global push. Eric Drexler effectively generated public interest in nanotechnology well before it became a subject of general discussion.

辑们着手工作的起点。这本书最重要的文字编辑是我的妻子 Barbara Whitesides,她对一些细节的关注极大地提高了这本书写作上的方方面面,包括从比喻的选择到语法的细节问题。她有着非凡的语言组织能力,并指出我认为可以却并非如此的想法。特别是在我使用代词"that"的时候,她常常会告诉我言不达意。正是她在文字编辑方面所做的努力,让一些我自己都无法继续读下去的文字变得可以容忍。Susan Wallace Boehmer 是哈佛大学出版社的一名编辑,是她让这本书看起来简洁明了,给我拿主意,最后使这本书趋于完美。她不但帮助我对比喻进行延伸,还对我的一些习惯性的书写错误不时纠正,以及不让我的文字单调乏味。Becky Ward 高效、优雅、熟练地将复杂的事物变得简单易行。

我的小儿子 Ben Whitesides 用他对词句的欣赏能力帮我添加了许多让人难忘的短句; 而大儿子 George T. Whitesides 则与 Ben 一起努力将我打造成一位称职的父亲。充当一位父亲与科学写作看起来似乎毫无关联,但事实并非如此。作为这个世界上最奇妙的工作,为人父母最为重要的是必须能够向充满好奇心的孩子们以任意的顺序解释世界上的任何事情。如果你自己还搞不清楚, 怎么能向别人解释清楚呢? George 和 Ben, Barbara, Felice, Mara, 以及我的学生们都是我的老师,作为他们的学生,我十分感激他们。

还有许多人,包括对纳米科学和我的教育事业作出贡献的科学家和其他人,都值得我在这里特别地感谢。Heine Rohrer 和 Gert Binning 发明了扫描探针显微镜从而开创了纳米领域;Cal Quate 最早向我解释了什么是纳米科学, 帮助我理解这一领域的机遇与挑战;Hank Smith 教会了我从现实出发去思考纳米刻蚀和纳米电子学。在这一领域的发展初期,作为一位不遗余力的倡导者,Mike Rocco 在华盛顿的努力对纳米科学在全球范围内的发展起到了重要的推动作用;Eric Drexler 让纳米技术在成为一个被广泛讨论的主题之前有效地赢得了公众的极大关注。Kavli 基金会对纳米科学的集中资助也推动其不断发展。

来自 Felice C. Frankel(原书作者之一)的感言:我十分感谢 Alfred P. Sloan 基金会对我写这本书的部分资助。我也感谢书中一些图片的贡献者以及他们允许我对其中一些图片进行修改或处理。我做这样的处理有时候是为了能更直观地展现图片的含义,有时候仅仅是为了让它变得更为有趣。这些贡献者的名字会出现在图片索引与简介部分,我希望读者们能多加留意。

Angela DePace,Rebecca Perry 和 Rosalind Reid 既是我的朋友也是我的同事,他们常常要求(甚至是逼迫)我用形象化的表达方式去思考科学,而这些方式我之前从未想象过。更为重要的是,她们一直强调我必须运用严谨的理性思维,有时候会直言不讳问我是怎么确定我的一些独特观点是正确的。Rebecca Ward 也对我起了同样的作用, 我在这里对她们一并表示感谢。Michael Halle 是我灵感的源泉,他总能了解我所想要知道的东西,并且总能让我追寻自己新的足迹。

与 Image Meaning (IM, www.imageandmeaning.org) 研讨会的众多参与者的交谈让我确信用图像的表达方式来阐释科学问题是非常有必要的。这场研讨会对我以及数百名的参与者来讲都是非常宝贵的经验,它让我们认识到用图像化的表达方式来阐释科学问题是一个巨大的挑战,同时也会对一个人如何去认知科学有着深远的影响。我非常感谢 Ruth Goodman 作为这个研讨会的主持人所作出的杰出贡献,我希望我们的努力会继续下去。

最优质和理智的监督者往往是我的学生们,是他们的存在让我一直坚持着自己正直的品质。幸运的是,我可以通过 NSF-Funded Picturing to Learn 这个项目(www.picturingtolearn.org)和学生们交换意见。这个项目的不断成功主要得益于项目主管 Rebecca Rosenberg 的辛勤工作。

我要特别感谢 Tim Jones,他是哈佛大学出版社的一名卓越的设计师。无形之中他出色地做到了将图片和文字巧妙地结合在了一起。同时我还要感谢 Eric Mulder,许多细节都由他最终敲定。

当然,最深的谢意和爱应该送给我的家人,他们一直都支持我致力于用图像来表达科学思想,包括 Matthew, Laura 和 Michael,以及刚刚出生的第三代 Yosi。我想不会有其他人比我更幸运了!

The Kavli Foundation's focused support for nanoscience has also helped it to grow.

From FCF: I am grateful to the Alfred P. Sloan Foundation for partial funding of my efforts for this book. I also want to thank the contributors of some of the images in the book and for their generosity in permitting me to "tweak" a few, sometimes for clarity, sometimes just for fun. Their names appear in the visual index, and I encourage readers to take note.

Angela DePace, Rebecca Perry, and Rosalind Reid are friends and colleagues who have nudged—no, *pushed*—me to think about science and its visual expression in ways I would never have imagined. But more important, they have insisted that I exercise intellectual rigor, sometimes not-so-subtly asking how sure I was of the accuracy of my particular perspective. Rebecca Ward has done the same, and I am grateful to them all. Michael Halle has been an inspiration—always "getting" what I needed to understand and always leading me to a host of relevant new tracks.

Conversations with the many participants in the Image and Meaning workshops (IM), www.imageandmeaning.org, have confirmed the need for the science community to understand the potential of visual explanations to clarify and explain. The workshops proved to be an invaluable experience for me and the hundreds of participants who walked away knowing well that the challenges in visually representing science can have a profound effect on how one *thinks* about science. I am most grateful to Ruth Goodman for her outstanding accomplishment as project manager of the IM workshops, and I hope our efforts will continue.

The very best of the intellectual police have always been the students. They keep me honest. I am lucky to continue our dialogues through the NSF-Funded Picturing to Learn program, www.picturingtolearn.org, whose ongoing success is mostly attributable to the hard work of its project manager, Rebecca Rosenberg.

Special gratitude goes to Tim Jones, our remarkable designer at Harvard University Press. Tim somehow brilliantly created an intelligent balance between the images and prose. My thanks also to Eric Mulder, who finalized the many details.

And always, my deepest gratitude and love to my family for indulging my obsession with the visual expression of science: Matthew and Laura, Michael, and now my amazing new grandchild, Yosi. How lucky can one person be?

译 后 记

很高兴能把《见微知著:纳米科学》这本书带给国内的广大读者。该书是三代纳米人共同努力的结晶:George Whitesides教授是夏幼南和秦冬在哈佛大学求学时的博士及博士后导师,而曾杰和王斅则分别是夏幼南在圣路易斯华盛顿大学的博士后及佐治亚理工学院的访问博士生。

这是怎样的一本书呢?英文版原著《No Small Matter: Science on the Nanoscale》于2009年由哈佛大学出版社出版。在Whitesides教授的笔下,纳米世界里神秘的原理和现象都逐一被生活中常见的事物与过程所代替,如拨云见日般把深奥难懂的微观世界变得异常的生动活泼、通俗易懂。散文般的文风时而欢畅如奔腾的溪流,时而动人似绕梁不绝的音律,同时充满了幽默诙谐和优雅神奇,处处折射出一位世界顶级科学家的睿智和博学。这本书在美国的读者群非常广,从中学生到大学生,再到那些为了启蒙小孩子的家长,甚至有一些培训机构也将它作为英语写作教材。

2012年7月,Whitesides教授来亚特兰大开会,与我们(夏幼南和秦冬)共进晚餐时表达了要把这本书在中国出版的意愿,这与我们的想法不谋而合。在美国讲授与纳米相关的课程时,我们就经常把这本书推荐给学生作为课外读物,以激发他们对纳米科学更浓厚的兴趣,同时也启发他们如何用简练的语言描述深奥的事物与复杂的过程。其实,早在这次见面之前,我们就已经把这本书的英文原版推荐给一些预备来美国读大学本科的国内中学生了,他们的反响十分强烈,大多觉得从中受益匪浅。起初,Whitesides教授只是希望把这本书译成中文版在国内发行,但原著中的许多英文原句实在是太经典了,那极强的感染力常常会让人在阅读的过程中情不自禁地陷入沉思。如果不能把这些既妙不可言又饱含哲理的英文原句呈现给广大读者,那实在是太遗憾了!经过跟Whitesides教授及哈佛出版社进一步磋商,最终,我们被授权翻译并获准在中国发行这本书的中英对照双语版。

紧接下来,曾杰(当时已作为"青年千人计划"获得者之一回到中国科学技术大学任教)和王斅便马不停蹄地开始了对原著的翻译工作。曾负责绪论及前面的30个章节,而王则负责剩余部分。截至2012年底,历时三个月,初稿便基本成型。但是,这只是万里长征的第一步!从来年的1月份起,一遍又一遍的修改工作便开始了。大至段落句子,小至符号空格,这种字斟句酌的修正一直持续到12月份。译得越深入、越能感受到这本书的震撼,而参与其中的也绝不仅仅只是曾杰和王斅了,夏和秦自不必说,还有国家纳米中心的蒋兴宇研究员夫妇、上海交通大学的邓涛教授、美国路易斯安那州立大学的王颖教授及夏幼南课题组的赵欣等人。或许是师出同门的缘故,在这为期一年多的修改工作中,大家的思维从未停止过碰撞,在办公室、家里,甚至是车站和机场的休息室,只要灵感的火花闪现,必定会在整个团队形成"星星之火,可以燎原"之势。交稿以后,出版社的责任编辑又以"信、达、雅"为准则,不断提高中文版的准确性与可读性。在描述波粒二象性时,Whitesides教授借用了一首简洁的现代诗来类比量子力学中的德布罗意方程之美。初稿中我们把这首诗译为:"你适合我,如钩入眼,一枚鱼钩,一只大眼。"虽然读起来还算押韵,但总觉得在意境上跟英文的诗还有一定差距。经过与责任编辑的几番讨论,最终改为:"你融入我,如钩入眼,一枚鱼钩,睁大的眼。"

一字一词皆千锤百炼,书名《见微知著:纳米科学》的由来也是颇具波折、不遑多让。早在翻译尚未启动之时,就有夏幼南充满禅意的《小非小,小之道》,曾杰的《三千微尘,三千世界》《纳米科学之见微知著》,亦有蒋兴宇夫妇那郑重其事的《不可小觑——纳米之美》……思维的交战如此激烈,以至在向出版社交稿后书名还是迟迟未定,到后来甚至连出版社的编辑也参与进来出了不少好点子。孰料造化弄人,当秦冬课题组的杨银同学再次提议《见微知著》时,才引起大家对这个书名的共鸣。这个最初的提议最终赢得了大家的认可,在改动了关键词的顺序之后,书名最终定为《见微知著:纳米科学》。"见微知著"取自《韩非子》,意指看到事物微观的细节,就会知道其在宏观上的特性与重要性。在这里用它来比喻以小见大,隐射纳米科学,即极小的东西却用途非凡。

千里马即出,伯乐安在?起初我们想将这本书融入"科学出版社"这个大家庭,但在沟通时方知,正是由于本书的"科学"性质,使得她必须进入"科学丛书",削成和丛书内其他书目一样的脸。此事让我们纠结了很久,因为这本书被大家视若珍宝,在我们心里,她是独一无二的。恰在此时,中国科学技术大学出版社的选题编辑向我们伸出了橄榄枝。中国科大,对于我们这个团队的很多人来说都是母校,像母亲对子女的宠溺一般,中国科大出版社也满足了我们的种种需求。

中国科大出版社负责本书出版工作的是一个平均年龄不到30岁的年轻团队。可以说,过于"吹毛求疵"的我们的确让他们费了不少神。光是在封面设计上,他们就前前后后给出了六个方案。我们从色彩、线条、表现手法等各个方面考量,最终选择了微流通道中的层流(Whitesides教授研究小组发明之一)作为封面,同时把Frankel(知名摄影家,原著作者之一)特别钟情的量子苹果图略加修改后放在了封底。另外,秦冬和她课题组的孙晓骏同学对封面设计、图像色彩和字体选择及安排的一体化,提出独到见解,将整体效应体现出英文版姐妹篇的精髓。希望大家的努力带给读者的不仅仅是视觉上的冲击,更多的是阅读的乐趣。

最后想说的是,中国科学院的白春礼院长在百忙之中特地为本书作了序,中国科学技术大学的侯建国校长也从未停止过对本书翻译和出版的关注与支持。在此我们谨代表原著作者把深深的敬意与谢意献给这两位中国纳米科学领域的带头人,非常感谢你们!也谢谢关注纳米科学的各位读者!

曾杰|王斅|秦冬|夏幼南

二零一四年三月

VISUAL INDEX

All images by Felice C. Frankel unless otherwise indicated.

图片索引与简介

除非特别说明，所有图像均由 Felice C. Frankel 制作。

1. Santa Maria PAGE 12

This clover-leaf map is a woodcut made in 1581 by Heinrich Bünting in Magdeburg. Jerusalem is in the center, surrounded by Europe, Asia, and Africa. The ship *Santa Maria* is reversed to correspond with the text. The image is from the Norman B. Leventhal Map Center at the Boston Public Library.

1. 探索新大陆(第12页)

这幅三叶草地图是 Heirich Bünting 于1581年在海德堡制作的一幅木雕。欧洲、亚洲与非洲以三叶草的形状排列，耶路撒冷位于中心。为了与文字对应，圣玛利亚号行驶的方向被反转了180度。图片来源于波士顿公共图书馆的 Norman B. Leventhal 地图中心。

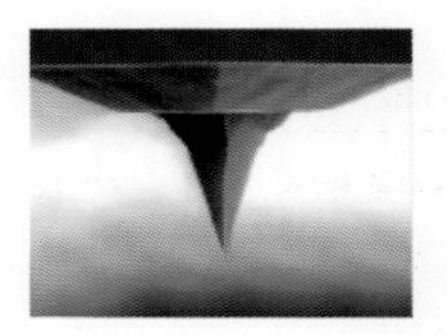

2. Feeling Is Seeing PAGE 15

This scanning electron microscope image of an atomic force microscope tip was generated with the help of Professor Christopher Love, MIT, and then digitally colored.

The field of pins in this image is a child's toy in which a rack of pins slides through a supporting sheet. Here, pieces of cardboard in contact with the back side of the array have pushed the pins forward, and caused them to replicate these shapes on the front side. The pins are about a millimeter in diameter.

2. 盲人摸象?(第15页)

这张原子力显微镜探针的扫描电子显微镜照片是在麻省理工学院的 Christopher Love 教授的帮助下拍摄完成的，后来用数码技术将其着色。

附图是一种儿童玩具：一个布满了可上下滑动的大头针的盘子。如果下方有一个物体向上挤压这些大头针，就可以在大头针上端的阵列上复显出该物体的形貌。这种大头针的直径在1毫米左右。

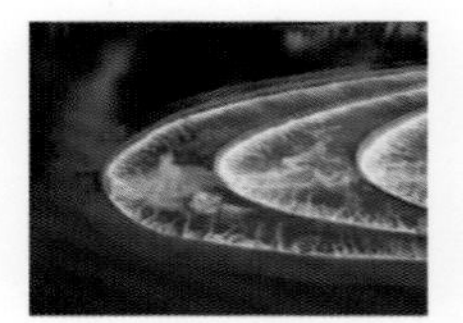

3. Quantum Cascades PAGE 18

This picture is of Pulteney Weir, on the River Avon, at Bath, in England. It was built in 1975 to reduce the risk of flooding.

3. 量子跃迁(第18页)

这张图片描绘的是位于英格兰巴斯市埃文河上的普尔特尼大坝，建成于1975年，其主要功能是防洪。

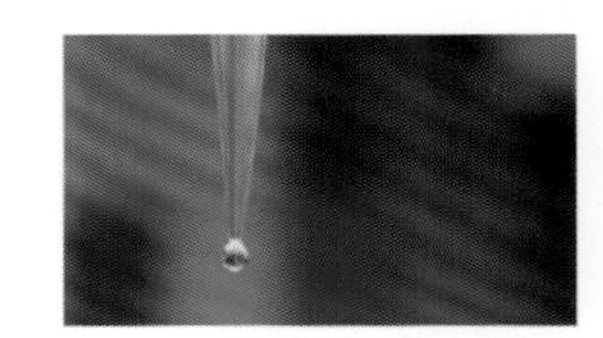

4. Water PAGE 21

The drop of water is suspended from a clear plastic pipette tip; only the sphere on the tip of the pipette is water. The length of the tip is approximately 3 centimeters.

4. 水(第21页)

这滴水珠悬挂在透明的塑料滴管尖端，呈球形。滴管尖端的长度大约有3厘米。

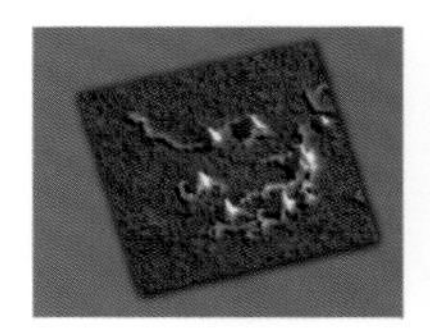

5. Single Molecules PAGE 22

The original AFM image of individual molecules of a diblock copolymer was obtained by C. Tang, C. J. Hawker, et al. in the Materials Research Laboratory at the University of California at Santa Barbara. J. Pyun et al., "Synthesis and Direct Visualization of Block Copolymers Composed of Different Macromolecular Architectures," *Macromolecules* 38 (2005): 2674－2685.

5. 单个分子(第22页)

这张双嵌段共聚物单体的原子力显微照片是由美国圣巴巴拉市加利福尼亚大学物质研究实验室的 C. Tang 和 C. J. Hawker 等人拍摄的。见：Pyun J, et al. 2005. Synthesis and Direct Visualization of Block Copolymers Composed of Different Macromolecular Architectures. Macromolecules 38: 2674－2685.

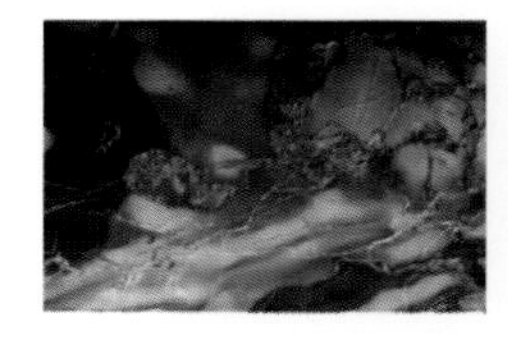

6. Cracks PAGE 25

This image of a colony of lichen growing on weathered rock was taken at Naumkeag, Prospect Hill Road, Stockbridge, Massachusetts.

6. 裂纹(第25页)

这张生长在风化岩上的地衣拍摄于和平天堂公园，该公园位于马萨诸塞州斯托克布里奇市的景山路。

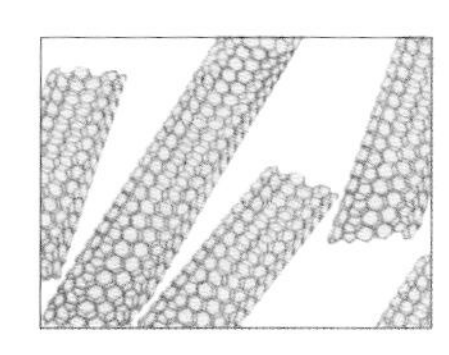

7. Nanotubes PAGE 26

In reality, single-walled carbon nanotubes can have diameters down to about 2 nanometers. Pages 158 - 159 explain how this photographic illustration was made.

7. 纳米管(第26页)

实际上,单层碳纳米管的最小直径可以达到约2纳米。第158 - 159页详细介绍了这张插图的由来。

8. Vibrating Viola String PAGE 30

This "flopped" image shows an old viola with one string vibrating.

8. 振动的琴弦(第30页)

这张看似很差的图展示的是一个古老的提琴,其中一根琴弦正在振动着。

9. Prism and Refraction PAGE 33

White light separates into colors on passage through this prism. An explanation of how the image was made can be found on pages 160 - 161.

9. 棱镜和折射(第33页)

白光通过棱镜后色散成彩色光带,这种现象又是怎么产生的呢?我们可以在第160 - 161页上找到答案。

10. Duality PAGE 34

The poem "You Fit into Me" is reproduced by permission of Margaret Atwood.

10. 波粒二象性(第34页)

这首"你融入我"的诗是经玛格丽特·阿特伍德歌授权使用的。

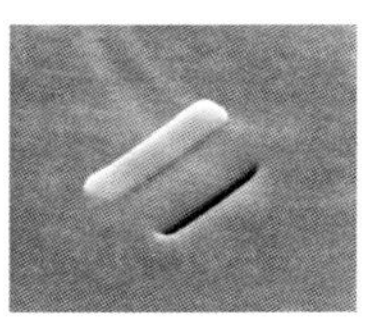

11. Interference PAGE 37

This experiment is a nanoscale recapitulation of the classic two-slit experiment demonstrating interference. The original experiment used light, while this one used electrons. Other experiments with slits of different geometries have actually demonstrated diffraction and interference from molecules passing through slits. Original images by S. Frabboni and G. C. Gazzadi, National Research Center on NanoStructures and BioSystems at Surfaces, Modena, Italy. S. Frabboni et al., "Nanofabrication and the Realization of Feynman's Two-Slit Experiment," *Applied Physics Letters*, 93 (2008): DOI 073108.

11. 波的干涉(第37页)

这个实验演示的是纳米尺度上的双缝干涉。与传统的干涉实验不同之处在于,这个实验用的是电子而非光。采用其他形状的狭缝,人们亦已观察到分子的干涉现象及其衍射条纹。图片由S. Frabboni和G. C. Gazzadi(意大利摩德纳表面纳米结构和生物系统国家研究中心)提供。见:Frabboni S, et al. 2008. Nanofabrication and the Realization of Feynman's Two-Slit Experiment. Applied Physics Letters, 93: DOI 073108.

12. Quantum Apple PAGE 38

An explanation of how this image was made is found on pages 154 - 155.

12. 量子苹果(第38页)

第154 - 155页介绍了这张照片的拍摄方式。

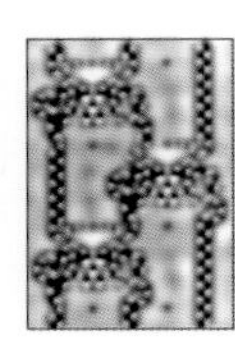

13. Molecular Dominoes PAGE 41

This image shows an array of molecules of carbon monoxide on a surface, placed with an AFM tip. The initial structure is a three-input sorter. Pushing the molecules at the top of the array imitates a cascade that propagates down the array. The end result is a computation of the digital logic functions AND, OR, and MAJORITY. Image by Don Eigler, IBM Research Division. A. J. Heinrich et al., "Molecule Cascades," *Science* 298 (2002): 1381 – 1387.

13. 分子多米诺(第41页)

图片展示的是采用原子力显微镜的探针看到的在固体表面上排列起来的一氧化碳分子。最初的结构是一个三输入端的分选器。推倒阵列最上端的分子,其余的也会随之倒下。最终的结果就像是数学中的逻辑函数"与"、"或"和"多数决定"的计算结果。图片由Don Eigler (IBM研究部)提供。见:Heinrich A J, et al. 2002. Molecule Cascades. Science, 298: 1381 – 1387.

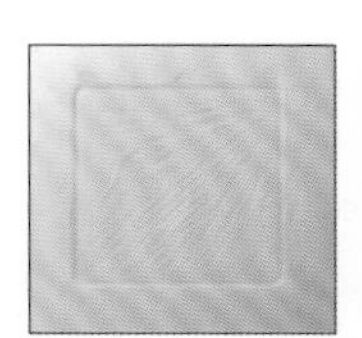

14. The Cell in Silhouette PAGE 42

This experiment was designed to understand the directional control of extension of lamellipodia by constraining the shape of the cell. This constraint was accomplished by allowing the cell to attach to a pattern of "squares" formed using a type of molecule that allowed cellular attachment, on a surface otherwise covered with molecules that prevented attachment. Original fluorescent image by Cliff Brangwynne and Don Ingber, Children's Hospital, Boston. K. K. Parker et al., "Directional Control of Lamellipodia Extension by Constraining Cell Shape and Orienting Cell Tractional Forces," *FASEB Journal* 16 (2002): 1195 – 1204.

14. 细胞肖像图(第42页)

这个实验的目的是理解如何通过约束细胞轮廓来定向控制板状伪足的延伸。在一个抗细胞吸附的表面上,将细胞能附着的某些分子排列在一个方形区域内,这样细胞就只能有选择性地附着在该区域,从而实现细胞的约束生长。原始的荧光照片由Cliff Brangwynne和Don Ingber (波士顿儿童医院)提供。见:Parker K K, et al. 2002. Directional Control of Lamellipodia Extension by Constraining Cell Shape and Orienting Cell Tractional Forces. FASEB Journal, 16: 1195 – 1204.

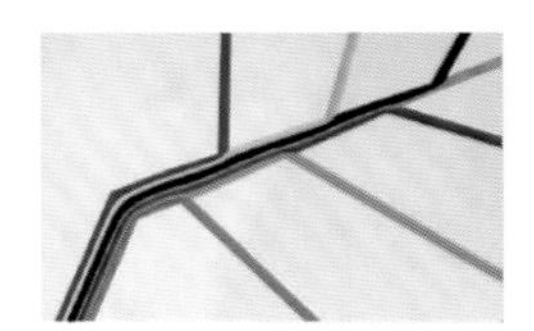

15. Laminar Flow PAGE 45

Liquids flowing in small channels (here, approximately 200 microns across) mix only by diffusion between the streams, not by formation of eddies or more complex processes. (Mixing does finally occur, by this diffusion, but the process is slow, and requires that the adjacent streams be thin.) In this image, each inlet stream feeds an aqueous solution of a colored dye into the system; the fluids exit (lower left) as a series of parallel, still unmixed streams. P. J. A. Kenis et al., "Microfabrication inside Capillaries Using Multiphase Laminar Flow Patterning," *Science* 285 (1999): 83 – 85.

15. 层流(第45页)

当两股在微通道(在此图中,大约200微米宽)中流动的水层相遇时,它们通过扩散来缓慢地混合,而不会有涡流或更复杂的过程产生(它们最终是会混合在一起,但速度十分缓慢,只适用于接触的水层都很薄的情况)。这张图片中,从不同微通道流入的是溶解了不同颜色染料的水溶液,当它们从左下端流出时,它们仍旧是一系列平行水层,并没有混在一起。见:Keinis P J A, et al. 1999. Microfabrication inside Capillaries Using Multiphase Laminar Flow Patterning. Science, 285:83 – 85.

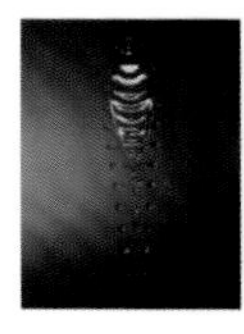

16. The Wet Fantastic PAGE 46

This structure formed when two jets of water were allowed to collide at an angle of somewhat more than 90°. The image shows one jet clearly; the water is going into the plane of the paper. This jet obscures the second, which is only partly visible, with the water coming out of the plane. Image by John W. M. Bush, MIT. A. E. Hasha and J. W. M. Bush, "Fluid Fishbones: Gallery of Fluid Motion," *Physics of Fluids* 14 (2002): S8.

16. 湿的奇谈(第46页)

当两股水以大于90度的接触角相碰撞时,就会形成图片中的景象。在这张图片中,我们较清晰地看到一股水流由纸外流向纸面,且这股水流挡住了第二股由纸面流向纸外的水流。图片由麻省理工学院的John W. M. Bush,提供。见:Hasha A E, Bush J W M. 2002. Fluid Fishbones: Gallery of Fluid Motion. Physics of Fluids, 14:S8.

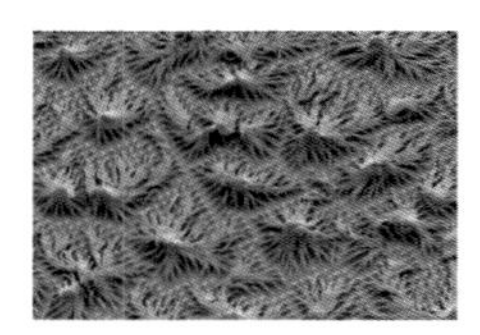

17. Fingers PAGE 49

The "posts" in this image are polymeric replicas, fabricated using nanomolding, of posts originally fabricated in silicon by electron beam lithography and directional etching. The structures formed when a suspension of polystyrene microspheres in water was placed on top of the array of posts and allowed to drain off and evaporate. Wetting of the posts, and capillarity, pulled these relatively flexible structures toward the spheres in the stage of evaporation in which the water film was only a few hundred nanometers thick. Image by Joanna Aizenberg, Harvard University. B. Pokroy et al., "Self-Organization of a Mesoscale Bristle into Ordered, Hierarchical Helical Assemblies," *Science* 323 (2009): 237 – 240.

17. "手指"(第49页)

图片中的这些聚合物"短棒"是用模板复制而成的。而模板则通过电子束与化学刻蚀在硅片表面产生。当一滴含有聚苯乙烯微球的水滴放在这些短棒阵列的基底上时,水开始挥发而变薄,随之产生的毛细作用力牵引着这些短棒伸向微球,最终得到图片中的结构。见:Aizenberg J, Pokroy B, et al. 2009. Self-Organization of a Mesoscale Bristle into Ordered, Hierarchical Helical Assemblies. Science, 323: 237 – 240.

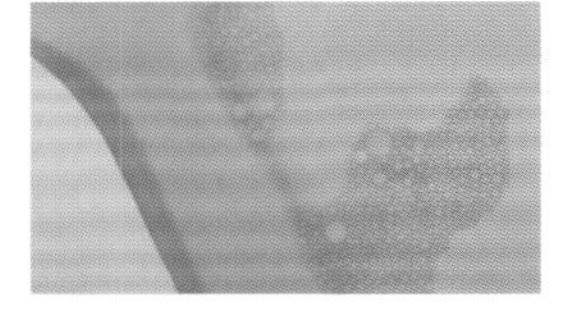

18. Soap Bubbles PAGE 53

These bubbles were formed in tap water, using household dishwashing detergent, and photographed on a green glass platter.

18. 肥皂泡(第53页)

这些肥皂泡沫是用日常生活中的洗洁精和自来水形成的,拍摄背景是一个绿色玻璃盘。

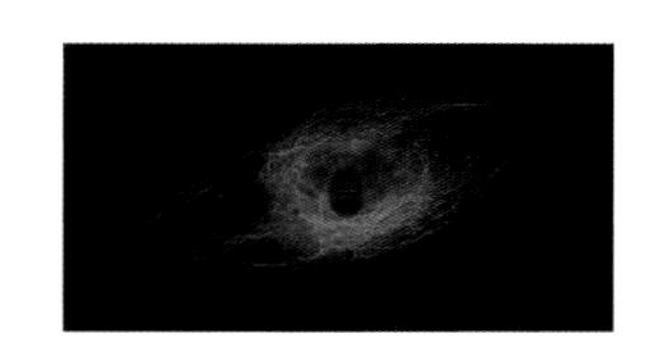

19. The Cell as Circus PAGE 54

The green filaments are images taken by fluorescence microscopy of green fluorescent protein (GFP) microtubules in a living bovine capillary endothelial cell. In this experiment, the gene for GFP (originally obtained from a jellyfish) was fused with the gene for the protein making up the microtubules, and the cell synthesized the fusion protein as it needed to form microfilaments. Image by Keiji Naruse and Don Ingber, Harvard Medical School and Children's Hospital, Boston.

19. 细胞与马戏团(第54页)

图片中绿色丝状物是通过荧光显微镜拍摄的牛毛细血管内皮活体细胞体内的绿色荧光蛋白(GFP)微管。实验过程中,产生GFP的基因(最初从海蜇中提取的)与产生微管蛋白的基因混合。当细胞需要微管时,便会合成出融合的蛋白质。图片由Keiji Naruse和Don Ingber(哈佛医学院和波士顿儿童医院)提供。

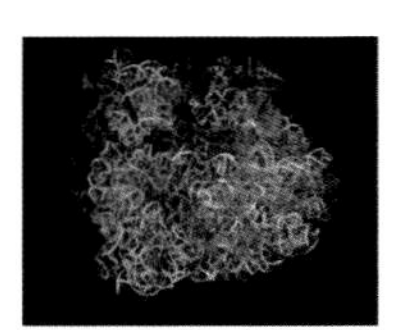

20. Ribosome PAGE 57

This image is a computer construction of a complex between a part of the ribosome (the 70S component) and a tRNA. It suggests some of the enormous complexity of this structure. Image by Harry Noller. A. Korostelev et al., "Crystal Structure of a 70S Ribosome-tRNA Complex Reveals Functional Interactions and Rearrangements," *Cell* 126 (2006): 1065 – 1077.

20. 核糖体(第57页)

图片展示的是部分核糖体与tRNA相作用而形成的复合物的计算机模型,这种结构极其复杂。图片由Harry Noller提供。见:Korostelev A, et al. 2006. Crystal Structure of a 70S Ribonsome-tRNA Complex Reveals Functional Interactions and Rearrangements. Cell, 126: 1065 – 1077.

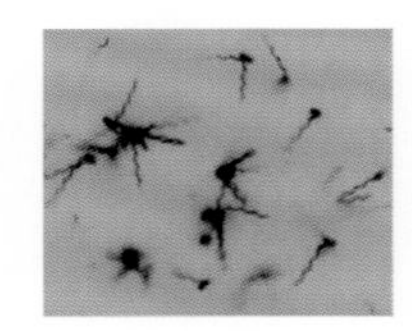

21. Bacterial Flagella PAGE 58

This is an image taken by high-resolution optical fluorescence microscopy. The body of a bacterium is approximately one micron in length. The original image, by Linda Turner and Howard Berg, was "inverted" in Adobe Photoshop. L. Turner et al., "Real-Time Imaging of Fluorescent Flagellar Filaments," *Journal of Bacteriology* 182 (2000): 2793 – 2801.

21. 细菌的鞭毛(第58页)

这是一张高分辨荧光显微镜照片。图中细菌大约有1微米长。图片由LindaTurner和Howard Berg提供,通过电脑软件对衬度反转而得到。见:Turner L,et al. 2000. Real-Time Imaging of Fluorescent Flagellar Filaments. Journal of Bacteriology, 182:2793 – 2801.

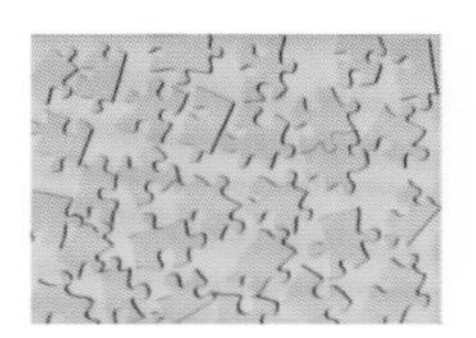

22. Life as a Jigsaw Puzzle PAGE 61

This image is of pieces of a jigsaw puzzle; it has been slightly manipulated to increase contrast.

22. 拼图般的生命(第61页)

图片中有很多块智力拼图,为了增加衬度,这些拼图被稍稍修改了。

23. As the Wheel Turns PAGE 62

This image is a technical tour de force. It provides a cross-sectional image of the structure that anchors the bacterial flagellar motor into the cell wall and membrane, and allows the motor proteins to generate the torque that drives rotation of the flagella. The motor proteins, and the proteins that translocate protons, do not survive the process used to isolate these structures, and are not included in the image. The nature of these protein structures is known from X-ray crystallographic experiments. This structure was obtained by rotational averaging of 101 separate images, taken in multiple orientations, in order to increase signal-to-noise and to emphasize common features. Image by David DeRosier, D. Thomas et al., "Structures of Bacterial Flagellar Motors from Two FliF-FliG Gene Fusion Mutants," *Journal of Bacteriology* 183 (2001): 6404 – 6412.

23. 转动的轮子(第62页)

图中所示真可谓是一项技术上的创举。描述了一个固定在细胞壁或细胞膜上的细菌鞭毛发动机的横截面,其中驱动蛋白质通过扭动来驱动鞭毛。驱动蛋白和质子传输蛋白在制样过程中没能被完整地保存下来,因此在图片中没有显示。这些蛋白质的结构已经用X射线晶体学方法获得。为了提高信噪比并突出相同点,科学家从不同角度拍摄,把101张不同的图片平均起来才得到这个图片。图片由David DeRosier提供。见:Thomas D, et al. 2001. Structures of Bacterial Flagellar Motors from Two FliF-FliG Gene Fusion Mutants. Journal of Bacteriology, 183:6404 – 6412.

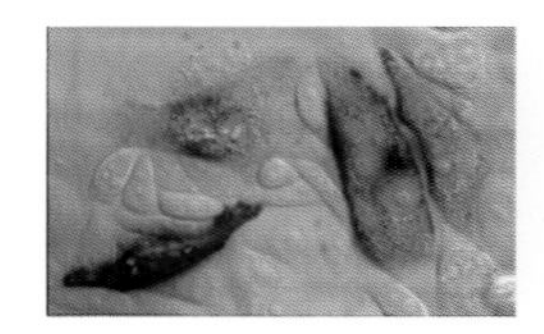

24. Quantum Dots and the Cell PAGE 65

"Transfection" is a collective name for a variety of processes that introduce DNA (and sometimes other biomolecules) into a mammalian cell, by opening pores in its membrane and encouraging uptake of the DNA by various means. In these images, fluorescent proteins were used to label cells that had been successfully transfected, and combinations of antibodies and quantum dots used to recognize specific protein receptors generated by the transfected cells. Original images by P. Zou and A. Ting, MIT. An explanation of how the final image was made can be found on pages 162 – 163. M. Howarth et al., "Monovalent, Reduced-size Quantum Dots for Imaging Receptors on Fixed Cells," *Nature Methods* 5 (2008): 397 – 399.

24. 量子点和细胞(第65页)

转染是外源DNA(或者其他生物分子)进入哺乳动物细胞这类过程的统称。转染的过程可以通过在细胞膜上穿孔,然后利用各种方法让DNA进入。图片中,荧光蛋白被用来标记成功转染的细胞,通过将抗体和量子点结合,来识别由转染过的细胞产生的特殊蛋白质接收基团。图片由麻省理工学院的P. Zou和A. Ting提供。关于这张图片的获取方式,请参考第162 – 163页。见:Howarth M, et al. 2008. Monovalent, Reduced-size Quantum Dots for Imageing Receptors on Fixed Cells. Nature Methods, 5: 397 – 399.

25. Sequencing DNA PAGE 66

Fifty years ago—in the dawn of DNA technology—sequencing was a technically difficult procedure involving analyses carried out by hand, and laborious development of sequence data by visual inspection of illuminated spots in thin gel films. Since then, the development of the technology for sequencing has been almost unbelievably rapid—substantially faster, by metrics of cost and size, than microelectronic technology. Now the entire process—from introduction of the sample to generation of the sequence—is entirely automated. For this image, DNA sequences were printed on two pieces of paper, which were then attached to a telephone book. The construction was then scanned on a flatbed scanner.

25. DNA 测序(第66页)

五十年前，在DNA技术发展的初级阶段，DNA测序还是一项很难操作的技术，因为人们必须通过人眼辨别凝胶薄膜中的亮点来进行数据的人工分析。此后，该项技术的发展速度让人难以想象，单以成本和数量来说甚至比微电子技术的发展还要快。如今，从测序的开始到完成，整个过程都已经实现了全自动化。在图片中，DNA的序列打印在两张纸上，然后贴在一本电话本上用一个平板扫描仪成像。

26. Molecular Recognition PAGE 69

This image is based on an X-ray crystal structure of the enzyme HMG-CoA reductase complexed with Simvastatin (a cholesterol-lowering drug). The crystal structure coordinates were obtained from the Protein Data Bank (http://www.rcsb.org/) using the PDB accession code 1HW9. The molecular model on which this image was based was produced by Demetri Moustakas, using the Chimera software package from the Computer Graphics Laboratory, University of California, San Francisco (supported by NIH P41 RR-01081). E. S. Istvan and J. Deisenhofer, "Structural Mechanism for Statin Inhibition of HMG-CoA Reductase," *Science* 292 (2001): 1160 - 1164.

26. 分子识别(第69页)

这张图片是用X射线单晶衍射测定的结构图，展示了HMG-CoA还原酶和辛伐他汀（一种降低胆固醇的药物）的复合物。它的晶体结构参数可以通过PDB加上1HW9的代码在蛋白质数据库中找到(http://www.rcsb.org/)。图中所示的分子模型是由Demetri Moustakas(计算图形实验室，加利福尼亚大学，旧金山)利用嵌合体软件包模拟得到的。见：Istvan E S, Deisenhofer J. 2001. Structural Mechanism for Stain Inhibition of HMG-CoA Reductase. Science, 292: 1160 - 1164.

27. Harvesting Light PAGE 70

Art by John Baldessari, *Beethoven's Trumpet (With Ear)*, Opus 127, 2007. Resin, fiberglass, bronze, aluminum, and electronics. 73 × 72 × 105 in. / 185.4× 182.8× 266.7 cm. Image courtesy of Marian Goodman Gallery, New York.

27. 捕获光(第70页)

作品名称为"贝多芬的小号"，作者：John Baldessari。作品由树脂、玻璃纤维、铜、铝和电子元件组成。尺寸：73 in × 72 in × 105 in/185.4 cm × 182.8 cm × 266.7 cm (1 in=2.54 cm)。图片由玛丽安古德曼画廊(纽约)提供。

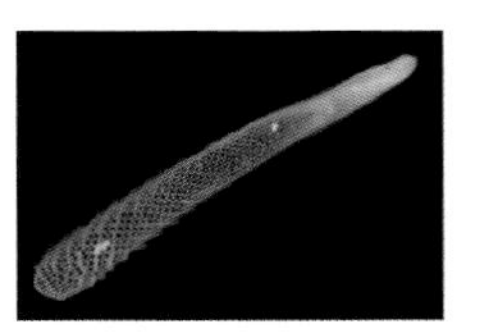

28. The Elegance of Simple Animals PAGE 73

The black and white scanning electron microscopic images in the collage were created by Joanna Aizenberg, Harvard University. The others were imaged on a flatbed scanner. All are of the mineral (silica nanospheres, in an organic matrix, but with compositions that vary with the position in the organism) skeleton of a sponge (Venus's Flower Basket, *Euplectella* sp.). The cells of the original, living organism have been removed. J. Aizenberg et al., "Skeleton of *Euplectella* sp.: Structural Hierarchy from the Nanoscale to the Macroscale," *Science* 299 (2005): 275 - 278.

28. 低等动物的优雅(第73页)

图中黑白的扫描电子显微镜照片由哈佛大学的Joanna Aizenberg拍摄，其余的是用平板扫描仪扫描得到的。所有这些结构都是矿物质(也就是埋在有机基质中的二氧化硅纳米球，其成分会随着在有机体中的位置变化而变化)海绵骨架(俗称"维纳斯花篮")。原始生命体的细胞已经被除去。见：Aizenberg J, et al. 2005. Skeleton of Euplectella sp.: Structural Hierarchy from the Nanoscale to the Macroscale. Science, 299: 275 - 278.

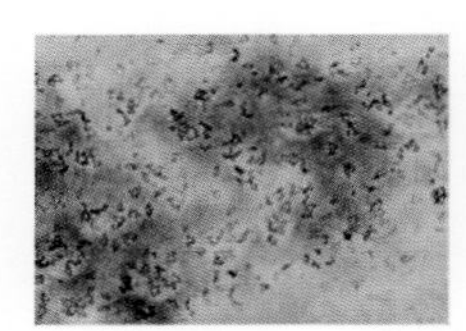

29. Antibodies PAGE 76

This transmission electron microscope image is of a mixture of monomeric IgGs (anti-DNP) and aggregates (mostly dimers and trimers). The aggregates formed when a solution of the IgGs was presented with divalent haptens, with the two ends capable of interacting with binding sites on two IgGs, but too short to bridge the binding sites on a single IgG. The original image by Jerry Yang first appeared in black and white, and was colorized for this book in Adobe Photoshop. J. Yang et al., "Self-Assembled Aggregates of IgGs as Templates for the Growth of Clusters of Gold Nanoparticles," *Angewandte Chemie International Edition* 43 (2004): 1555 – 1558.

29. 抗体(第76页)

这张透射电子显微镜图像展示的是IgG型抗体(anti-DNP)及其聚集物(主要为二聚体和三聚体)。二价的半抗原两端都能与IgG型抗体的结合位点相连接,但是由于尺寸的原因,这两端不能与同一个抗体的结合位点相结合。因此,将IgG溶液与这种半抗原混合就能获得抗体的聚集体。由Jerry Yang拍摄的原始图片是黑白的,这张插图是经过Photoshop软件着色后得到的。见:Yang J, et al. 2004. Self-Assembled Aggregates of IgGs as Templates for the Growth of Clusters of Gold Nanoparticles. Angewandte Chemie International Edition, 43: 1555 – 1558.

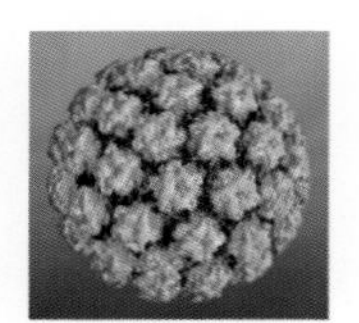

30. Virus PAGE 79

This image is a computer reconstruction of the capsid (the highly structured aggregate of proteins that forms the outer shell of a virus) of a human papillomavirus particle. This type of virus is of substantial medical interest, since papilloma viruses are associated with disorders ranging from cervical cancer to warts. Image by Yorgo Modis and Steve Harrison, Yale University. Y. Modis et al., "Atomic Model of the Papillomavirus Capsid," *EMBO Journal* 21 (2002): 4754 – 4762.

30. 病毒(第79页)

该图片是用计算机模拟出来的人类乳头状瘤病毒粒子的衣壳(展示了一种高度结构化的蛋白质分子聚集体)。研究表明,这种病毒与宫颈癌和疣都有密切关系,因此对乳头状瘤病毒的研究具有重要的医学价值。图片由耶鲁大学的Yorgo Modis和Steve Harrison提供。见:Modis Y, et al. 2002. Atomic Model of the Papillomavirus Capsid. EMBO Journal, 21: 4754 – 4762.

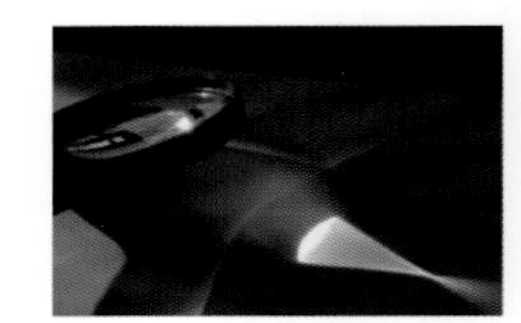

31. Writing with Light PAGE 83

The magnifying glass on the left produces these light forms, which are the basis for a number of studies in physics, and included in the fascinating field of caustics.

31. 光刻(第83页)

图片左端的放大镜产生的这些光学现象是许多物理学研究的基础,包括前景诱人的焦散领域。

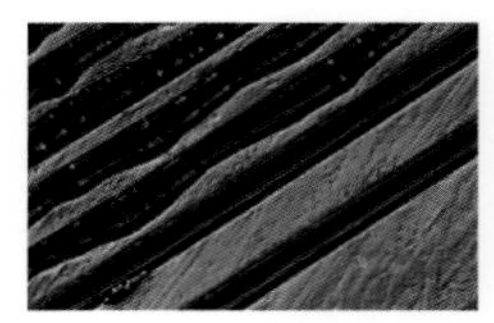

32. Eleanor Rigby PAGE 84

The Beatles' vinyl album Revolver yielded this image using a microscope with Nomarski Differential Contrast.

32.《埃莉诺·里格比》(第84页)

这是用诺马斯基微分相差显微镜拍摄到的披头士音乐专辑"Revolver"的黑胶唱片。

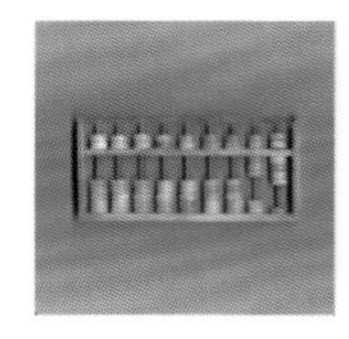

33. Abacus PAGE 87

There are several different systems used in abaci. This one is still often used in China. Its use is described in a number of sources, including http://www.ee.ryerson.ca/~elf/abacus/intro.html.

33. 算盘(第87页)

算盘有多种进制。图中的这种算盘在中国仍旧有人在使用。关于这种算盘的使用方法,请参考:http://www.ee.ryerson.ca/~elf/abacus/intro.html.

34. Counting on Two Fingers PAGE 88

This image illustrates binary addition. It's a very easy system to learn, although easier for computers to use than for people. Wikipedia has a clear explanation for those who might be interested in actually trying it: (http://en.wikipedia.org/wiki/Binary_numberal_system). The arrangement of glasses illustrates this addition:

001100
+010110
=100010

See pages 156 - 157 for an explanation of how the image was made.

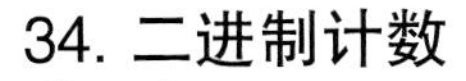

34. 二进制计数（第88页）

这张图片展示了二进制加法，这是一种很容易学的算法，尤其对计算机而言。对此算法感兴趣的人，可以参考维基百科中的详细阐述。玻璃杯的不同排列方法演示了这一算法：

001 100
+ 010 110
= 100 010

至于这个图像是如何拍摄的请参考书中第156 - 157页。

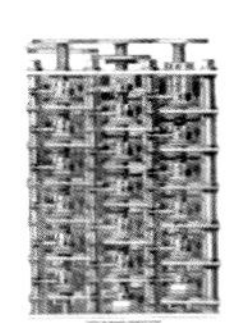

35. Babbage's Computing Engine PAGE 91

Image courtesy of the Science and Society Picture Library, London, England.

35. 巴贝奇的计算器（第91页）

图片由英国伦敦的科学和社会图片图书馆提供。

36. Computers as Waterworks PAGE 92

These pipes and valves are part of a water system in Cambridge, Massachusetts. The image was simplified by digitally deleting parts of a distracting background.

36. 供水系统般的计算机（第92页）

这些管道和阀门是马萨诸塞州坎布里奇市供水系统的一部分。该图片经过数码技术后期处理并删除部分背景后获得。

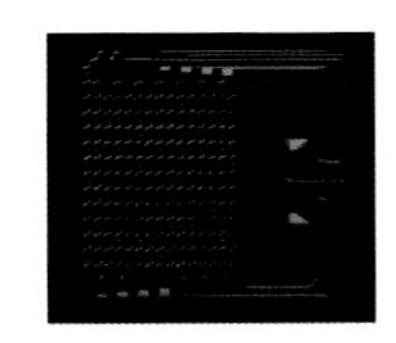

37. Microreactor PAGE 95

This microreactor was developed for the continuous synthesis of nanocrystals. Working in a microfluidic system (the channels are about 400 microns across) brings a greater measure of control over the system than can be achieved with macroscopic systems. B. K. H. Yen, "Microfluidic Reactors for the Synthesis of Nanocrystals," Ph.D. dissertation, MIT, Department of Chemistry, 2007. The journal *Lab on a Chip* describes many applications of this kind of device.

37. 微反应器（第95页）

这个微反应器技术是用来合成纳米晶体的。比起宏观体系，用微分流系统（通道约400微米宽），可以更有效地控制很多反应条件。见：Yen B K H. 2007. Microfluidic Reactors for the Synthsis of Nanocrystals, Ph. D. dissertation, MIT, Department of Chemistry.《Lab on Chip》杂志里有许多关于类似应用的报导。

38. Templating PAGE 97

The black and white image of the block copolymer, self-organized by epitaxial self-assembly, was generated by scanning electron microscopy. The period of the structures in the image is 48 nanometers. Image by Paul Nealey, University of Wisconsin. The larger image of chairs, set up in lines for the parents of graduating students, was taken at MIT. S. O. Kim et al., "Epitaxial Self-Assembly of Block Copolymers on Lithographically Defined Nanopatterned Substrates," *Nature* 424 (2003): 411 - 414.

38. 模板效应（第97页）

黑白图片显示的是嵌段共聚物经外延自组装排列后的扫描电子显微镜图像。图中每个重复单元约为48纳米宽。图片由威斯康星大学的Paul Nealey提供。附图中成排的椅子拍于麻省理工学院的一个毕业典礼。见：Kim S O, et al. 2003. Epitaxial Self-Assembly of Block Copolymers on Lithographically Defined Nanopatterned Substrates. Nature, 424: 411 - 414.

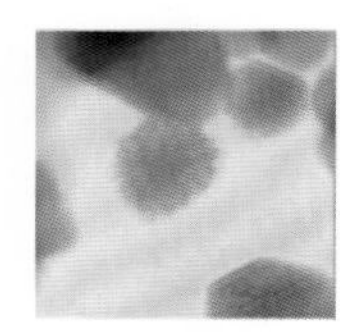

39. Catalyst Particles PAGE 101

This digitally-colored transmission electron micrograph shows particles of composition "Pt_3Co," whose purpose is to provide enhanced catalytic activity for electrochemical reduction of O_2 in various applications, such as fuel cells. The original image was in black and white, obtained at Oak Ridge National Laboratory (MIT), P. J. Ferreira (University of Texas, Austin), and L. F. Allard (ORNL). P. J. Ferreira and L. F. Allard, "Enhanced Activity for Oxygen Reduction Reaction on 'Pt_3Co' Nanoparticles: Direct Evidence of Percolated and Sandwich-Segregation Structures," *Journal of the American Chemical Society* 130 (2008): 13818 – 13819.

39. 催化剂颗粒(第101页)

图片是Pt_3Co纳米颗粒的透射电子显微镜图像。这种纳米颗粒对电化学氧气还原反应具有良好的催化活性，在燃料电池等领域具有广泛应用。原始图片在橡树岭国家实验室(ORNL)由Y. Shao-Horn(MIT), P. J. Ferreira(德克萨斯大学，奥斯汀)和L. F. Allard (ORNL)拍摄，原始图片是黑白的，这张彩图是经过计算机着色处理后所得。见：Ferreira P J, Allard L F. 2008. Enhanced Activity for Oxygen Reduction Reeaction on "Pt_3Co" Nanoparticles: Direct Evidence of Percolated and Sandwich-Segregation Structures. Journal of American Chemical Society, 130: 13818 – 13819.

40. Christmas-Tree Mixer PAGE 102

One of the major differences between fluid flows in small channels ("microfluidics") and in large channels is the absence of turbulent mixing. This device (nicknamed a "Christmas-tree mixer") makes it possible to mix two fluids flowing side by side in very thin (50 micron) streams using diffusion. That mixing allows the formation of gradients using appropriate microchannel designs. These gradient mixers are used in a range of experiments in cell biology. Image from the Whitesides laboratory, Harvard University. N. L. Jeon et al., "Generation of Solution and Surface Gradients Using Microfluidic Systems," *Langmuir* 16 (2000): 8311 – 8316.

40. 圣诞树般的混合器(第102页)

当流体在微小通道与大尺寸通道中流动时，其主要差别之一就是，微小通道中的流体不会发生涡流混合。这个装置(别名：圣诞树般的混合器)利用扩散作用，成功实现了将两股并行的流体在很窄的通道(50微米宽)混合在一起。通过合理的微通道设计，这种混合器能产生具有一系列浓度梯度的混合溶液。这种混合器在细胞生物学领域具有广泛应用。图片由哈佛大学的Whitesides课题组提供。见：Jeon N L, et al. 2000. Generation of Solution and Surface Gradients Using Microfluidic Systems. Langmuir, 16: 8311 – 8316.

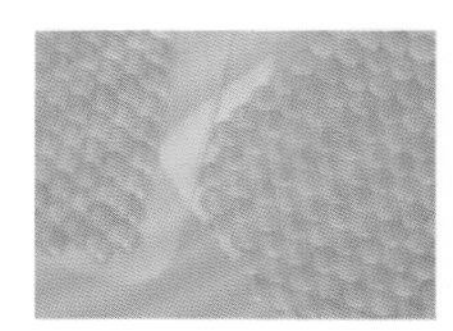

41. Self-Assembly PAGE 105

One-centimeter glass balls were poured onto two glass dishes with slightly curved surfaces, and tapped lightly. They "crystallized" immediately.

41. 自组装(第105页)

图片展示的是放置在两个底部轻微凹陷的玻璃盘子里的厘米大小的玻璃球。只需要轻轻敲打几下盘子的边缘，这些球立刻就会"结晶"成整齐的结构。

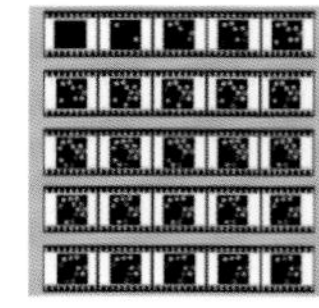

42. Synthetic Nose PAGE 106

Single images from a video supplied by Todd Dickinson and David Walt, Tufts University, were digitally "seamed" together to simulate a film-strip suggesting lapsing time. The video is an optical readout of a sensor reading organic vapors. T. A. Dickinson et al., "A Chemical-Detecting System Based on a Cross-Reactive Optical Sensor Array," *Nature* 382 (1996): 697 – 700.

42. 人造狗鼻子(第106页)

这一组图来自塔夫茨大学的Todd Dickinson和David Walt的视频。我们用电脑把它们串在一起以显示时间的变化。它们描述了一个有机材料传感器在识别气体过程中光输出信号的变化过程。见：Dickinson T A, et al. 1996. A Chemical-Detecting System Based on a Cross-Reactive Optical Sensor Array. Nature, 382: 697 – 700.

43. Millipede PAGE 109

The millipede was originally developed by IBM. One possible application was as a reader for information. This particular application may not be commercialized, but devices using multiple atomic force microscope or scanning tunneling microscope tips are being developed for a wide range of applications, from printing biochemical sensors to reading information. Image courtesy of IBM, Zurich.

43. 千足虫(第109页)

图片中这个类似于千足虫的装置最初是由IBM公司设计的，它可能会被用作信息读取工具。虽然这个应用大概很难商业化，但是多探针的原子力显微镜和扫描隧道显微镜正在被应用于多个领域，例如生化传感器的制作和信息读取。图片来源：IBM，苏黎世。

44. E-paper and the Book PAGE 110

This e-paper is the material that provides the page for current e-books. This sample was supplied by E-Ink Corporation, Cambridge, Massachusetts.

44. 电子纸和电子书(第110页)

电子书离不开图中所示的电子纸。这个样品由马萨诸塞州坎布里奇市的E-Ink公司提供。

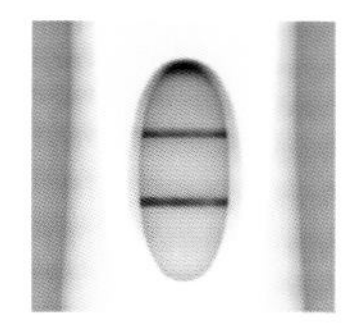

45. Lateral-Flow Assay as Crystal Ball PAGE 113

The positive pregnancy test imaged here was obtained using a commercial product that uses lateral flow immunoassay.

45. 决定命运的横向流动检测法(第113页)

图片中这个显示阳性的怀孕测试棒是基于横向流动免疫检测法的一个商业产品。

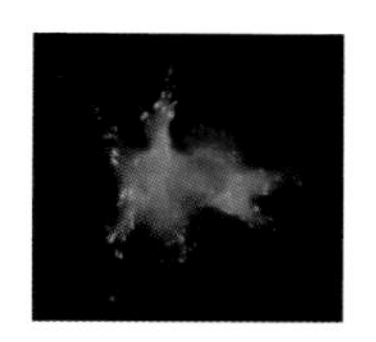

46. Testing Drugs in Cells PAGE 114

This image shows a mammalian cell cultured in a Petri dish, taken using fluorescence microscopy. This picture illustrates particularly clearly the points at which the cell attaches to the surface of the dish, and some features of its internal structure. Image by Julia Sero and Don Ingber, Harvard Medical School and Children's Hospital, Boston.

46. 用细胞来筛选药物(第114页)

这是一个在培养皿中得到的哺乳动物细胞的荧光显微相片。图片清晰地展示了细胞附着在培养皿表面的细节及其内部构造的一些特征。图片由Julia Sero和Don Ingber(哈佛医学院和波士顿儿童医院)提供。

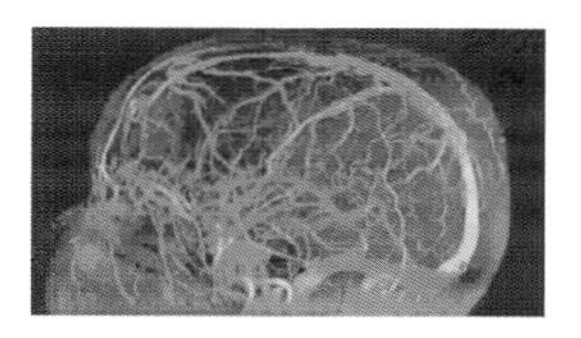

47. Cooling the Fevered Brain PAGE 117

Injection of magnetic nanoparticles (dextran-coated magnetite) into the blood stream enhances the contrast in magnetic resonance imaging between flowing blood and tissue, based on the ability of these particles to shorten the relaxation time of water molecules. The original black and white image was provided by Advanced Magnetics, Inc., Cambridge, Massachusetts, which also makes the magnetic particles. Colors were added in Adobe Photoshop.

47. 给发烧的大脑降温(第117页)

磁性粒子可以缩短水分子的弛豫时间，因此，在磁共振成像过程中，将磁性纳米粒子注入血液中就可以增强流动的血液与组织器官成像的衬度。原始照片由马萨诸塞州坎布里奇市的Advanced Magnetics有限公司（该公司生产磁性颗粒）提供，后来是通过Photoshop着色而成的。

48. Phantoms PAGE 120

This photograph is of the Tanner Fountain at Harvard University, during summer (when it sprays a cooling mist over the rocks—to the delight of children). The image was originally taken for a book about modern landscape architecture.

48. 幽灵(第120页)

这是一张哈佛大学泰纳喷泉的照片。夏天的时候,从喷泉中洒出的细水珠变成悬浮在石头上的雾霭,这一景象深受孩子们的喜爱。该相片原本是为一本关于现代园林风景的书所摄。

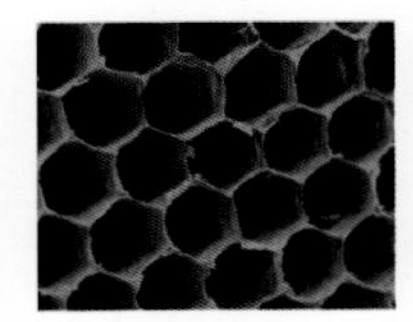

49. Privacy and the Nest PAGE 123

This detail of a hornet's nest (*Vespa crabro*) was photographed in the Department of Entomology, Harvard University Museum of Comparative Zoology.

49. 隐私与蜂窝(第123页)

这张马蜂窝(黄边胡蜂)的高清图拍摄于哈佛大学昆虫学系的比较动物学博物馆。

50. Soot and Health PAGE 124

The combined smoke from three lighted cigarettes gave this image.

50. 灰尘与健康(第124页)

图中所示是由三根香烟燃烧所产生的烟雾。

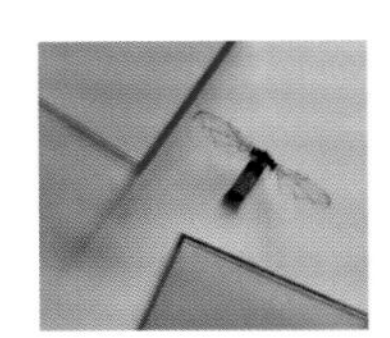

51. Robots PAGE 127

The small fly is part of a program by Rob Wood at Harvard to build microvehicles capable of autonomous flight without human intervention or control. The fly was supplied by Ben Finio, Katie Byl, and Jessica Shang of the Harvard Microrobotics Laboratory. R. J. Wood, "The First Takeoff of a Biologically Inspired At-Scale Robotic Insect," *IEEE Transactions in Robotics* 24 (2008): 341 - 347. The music box mechanism is designed to play "Jingle Bells."

51. 机器人(第127页)

这只小苍蝇是哈佛大学 Robert Wood 课题组研制的可无人操控自动飞行的微型飞行器。这个飞行器由 Ben Finio, Katie Byl 和 Jessica Shang (哈佛大学微型机器人实验室)提供。见:Wood R J. 2008. The First Takeoff of a Biologically Inspired At-Scale Robotic Insect. IEEE Transactions in Robotics, 24: 341 - 347。附图的音乐盒可以演奏乐曲《Jingle Bells》。

52. Fog PAGE 130

Photographed at John Deere Headquarters, Moline, Illinois, for a book about modern landscape architecture.

52. 雾(第130页)

图片拍摄于伊利诺伊州莫林市的 John Deere 总部,选自一部关于现代园林风景的书。

53. In Sickness and in Health PAGE 133

This *Bacillus subtilis* sample, growing on a gel medium in a Petri dish, was prepared by Hera Vlamakis in the laboratory of Roberto Kolter at the Harvard Medical School. The colony is about 2 centimeters across, and approximately a millimeter thick.

53. 疾病与健康(第133页)

图片中的这个枯草杆菌样品是由 Hera Vlamakis (Roberto Kolter 实验室,哈佛医学院)在凝胶介质上培养得到的。这个样品的截面积大约为 2 平方厘米,厚度大约是 1 毫米。

54. The Internet PAGE 136

Both images are computer representations of large sets of data. The upper visualization shows flight patterns of commercial aircraft: "Flight Patterns - Aaron Koblin 2009." The lower shows bit flows on the internet, and was prepared by Chris Harrison, Carnegie Mellon University.

54. 因特网(第136页)

这两张计算机产生的图像用来描述大量的数据。上面一幅由来往各地的飞机航线组成(图片选自《飞行图案》,2009,作者:Aaron Koblin)。下面一幅则由城市间网络上传输的信息构成(作者:Chris Harrison,卡内基梅隆大学)。

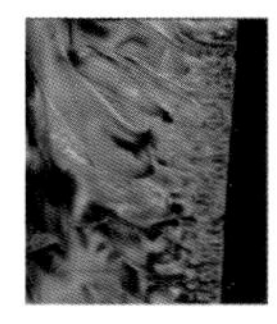

55. Reverse Osmosis Membrane PAGE 139

Reverse osmosis membranes are designed and fabricated to have an elaborate, asymmetrical structure by precipitation of polymer from a solution at an interface. The precipitation develops the network of pores that provide—on the front face—the capability for discriminating water from ions, and in the rest of the membrane the mechanical strength and channel system needed for the membrane to withstand the conditions in which it operates. This electron microscope image was originally in black and white, and was colorized using Adobe Photoshop. Original black and white image by A. Asatekin, MIT. A. Asatekin et al., "Anti-fouling Ultrafiltration Membranes Containing Polyacrylonitrile-*graft*-poly (ethylene oxide) Comb Copolymer Additives," *Journal of Membrane Science* 298 (2007): 136 - 146.

55. 反渗透膜(第139页)

反渗透膜是一种通过聚合物在界面上沉积而制成的具有独特的非对称结构的薄膜。沉积法有助于网络状孔道的形成,在膜表面的孔状结构能够有效地使水与盐离子分离,而其他部分则能增强机械强度,使膜能够承受操作过程中巨大的压力。原始的电子显微照片是黑白的,这张彩图是利用Photoshop着色而成的。原始图片来自MIT的A. Asatekin。见:Asatekin A, et al. 2007. Anti-fouling Ultrafiltration Membranes Containing Polyacrylonitrile-graft-poly (ethylene oxide) Comb Copolymer Additives. Journal of Membrane Science, 298: 136 - 146.

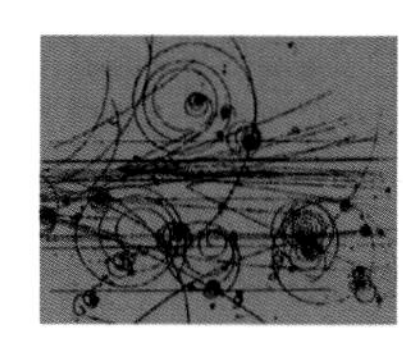

56. Nuclear Reactions PAGE 140

Trails of charged atomic particles moving through liquid hydrogen in a bubble chamber; tiny bubbles mark the paths energetic particles have taken. This colored image shows tracks produced by collisions of pions with hydrogen atoms. The curling shape reflects movement of charged particles (largely electrons) in an imposed magnetic field. The curvatures help to characterize the mass and charge of these particles. Original black and white image from CERN.

56. 核反应(第140页)

该图显示的是带电粒子在气泡室的液氢中运动的踪迹图,科学家们利用一连串的小气泡记录下了这些粒子的运动路径。这张图片显示的运动轨迹是π介子与氢原子碰撞后得到的。其曲线显示了在外加磁场的作用下带电粒子(大部分为电子)的运动形式。从曲线的曲率可以测量这些粒子的质量和电荷。原始图片来源:欧洲核子研究中心(CERN)。

57. Flame PAGE 143

This image is just what it seems: a candle (four inches wide) that would decorate a dinner table.

57. 火焰(第143页)

显而易见,这是一根装点餐桌的蜡烛(直径:4英寸)。

58. Fuel Cell PAGE 144

This membrane is a multilayer construction, designed in part to resist permeation by methanol, in a type of fuel cell that uses methanol (rather than hydrogen) as fuel. The piece of membrane shown here is several centimeters across, and approximately a millimeter thick. It comprises a number of different layers, each with different functions. Membrane supplied by A. Argun, MIT. A. Argun et al., "Highly Conductive, Methanol Resistant Polyelectrolyte Multilayers," *Advanced Materials* 20 (2008): 1539 - 1543.

58. 燃料电池(第144页)

图中的薄膜是为了防止甲醇（而非氢气）燃料电池中甲醇的透过而设计的多层结构薄膜。这张薄膜有几个平方厘米大小，厚度约1毫米，由多个具有不同功能的薄层组成。薄膜由麻省理工学院的A. Argun提供。见：Argun A, et al. 2008. Highly Conductive, Methanol Resistant Polyelectrolyte Multilayers. Advanced Materials, 20: 1539 - 1543.

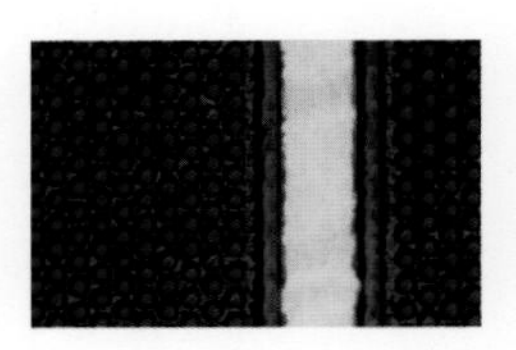

59. Solar Cell PAGE 147

This image is of a microscopic area of the top side of a silicon-wafer-based photovoltaic solar cell with a low-reflectivity surface consisting of a hexagonal array of pits. The surface appears blue-violet due to an antireflective coating of silicon nitride. Light that is absorbed in the wafer creates charge that moves to the surface to be collected by the narrow (<40 micron) silver conductor lines shown in the image as a gold-colored strip. Sample wafer supplied by 1366 Technologies, Lexington, Massachusetts.

59. 太阳能电池(第147页)

图片是硅基太阳能电池顶板的显微照片，它的表面由整齐排列的六边形凹槽组成，因而具有低的光反射率，由于镀了一层抗反射的氮化硅而呈现蓝紫色。硅片吸收太阳光所产生的电荷迁移到这个表面上，然后被极细的银导线（图中的金色带状部分）收集。样品由1366 Technologies（马萨诸塞州列克星敦市）提供。

60. Plants and Photosynthesis PAGE 148

Photograph taken at the Bloedel Reserve, Bainbridge Island, Washington, in a garden originally designed by Rich Haag.

60. 植物和光合作用(第148页)

照片拍摄于布洛德保护区（位于美国华盛顿州班布里奇岛），这个公园是由Rich Haag设计的。